Wireless Network Security

Wolfgang Osterhage
Goethe-Universität
Frankfurt, Germany

CRC Press
Taylor & Francis Group
Boca Raton   London   New York

CRC Press is an imprint of the
Taylor & Francis Group, an **informa** business

A SCIENCE PUBLISHERS BOOK

CRC Press
Taylor & Francis Group
6000 Broken Sound Parkway NW, Suite 300
Boca Raton, FL 33487-2742

Visit the Taylor & Francis Web site at
http://www.taylorandfrancis.com

and the CRC Press Web site at
http://www.crcpress.com

Preface

The first edition of "Wireless Network Security" was published, only five years ago, in 2012. However, looking at the pace of technological advancements and developments which have surpassed the state-of-the-art presented then, it was time to take these into consideration. The present edition comprises:

- New versions of the IEEE 802.11 standard
- The omission of the chapter on PDAs, since their functions have been absorbed by Smart Phones
- The technologies and services of Smart Phones
- New versions of Bluetooth up to V 5.0 and specific threats
- Near Field Communication
- Networks security in general
- Emergency Management.

Wolfgang Osterhage
February 2018

Contents

◇◇

Abbreviations

AAI   Authentication Algorithm Identification
ACL   Asynchronous Connectionless
AES   Advanced Encryption Standard
ANSI   American National Standard Institute
App   Application
ARP   Address Resolution Protocol
ASCII   American Standard Code for Information Interchange
AuC   Authentication Center
BCM   Business Continuity Management
BDA   Bluetooth Device Address
BES   BlackBerry Enterprise Server
BIA   Business Impact Analysis
BSC   Base Station Controller
BSS   Basic Service Set
BTS   Base Transceiver Station
Cal   Calendar
CCK   Complementary Code Keying
CD   Compact Disc
CEPT   Conference Europeéne des Administrations des Postes et
               Télècommunications
CF   CompactFlash
CRC   Cyclic Redundancy Check
CSMA   Carrier Sense Multiple Access
CSMA/CA   Carrier Sense Multiple Access/Collision Avoidance
CSMA/CD   Carrier Sense Multiple Access/Collision Detection
CTS   Clear to Send
DFS   Dynamic Frequency Selection
DHCP   Dynamic Host Configuration Protocol
DoS   Denial of Service
DSL   Digital Subscriber Line

DSSS	Direct Sequence Spread Spectrum
DUN	Dial-up Network
DVD	Digital Versatile Disc
EAP	Extensible Authentication Protocol
EIR	Equipment Identity Register
EMS	Enhanced Message Service
ESS	Extended Service Set
ETSI	European Telecommunications Standardisation Institution
EU	European Union
FCC	Federal Communications Commission
FHSS	Frequency Hopping Spread Spectrum
FMC	Fixed Mobile Convergence
FTP	File Transfer Profile
GAP	Generic Access Profile
GFSK	Gaussian Shift Keying
GHz	Gigahertz
GPRS	General Packet Radio Service
GPS	Global Positioning System
GSM	Global System for Mobile Communications
GUI	Graphical User Interface
HID	Human Interface Device Profile
HIPERLAN	High Performance Radio Local Network
HomeRF	Home Radio Frequency
HR-DSSS	High-Rate Direct Sequence Spread Spectrum
HLR	Home Location Register
HSCSD	High Speed Circuit Switched Data
HSDPA	High Speed Downlink Packet Access
HSP	Headset Profile
IBSS	Independent BSS
ICV	Integrity Check Value
ID	Identifier
IEEE	Institute of Electrical and Electronic Engineers
IMSI	International Mobile Subscriber Identity
IP	Internet Protocol
ISM	Industrial, Scientific, Medical
ISO	International Organisation for Standardisation
IT	Information Technology
ITU	International Telecommunications Union
IV	Initialisation Vector
JIS	Japanese Industrial Standard
Kbit/s	Kilobits per second
kHz	Kilohertz

km	Kilometre
L2CAP	Logical Link Control and Adaptation Protocol
LAN	Local Area Network
LLC	Logical Link Control
m	Metre
MAC	Medium Access Control
MAN	Metropolitan Area Network
MBit/s	Megabits per second
MDS	Mobile Data Service
MHz	Megahertz
MIMO	Multiple Input Multiple Output
MMS	Multimedia Messaging Service
MP3	Moving Picture Experts Group Layer-3 Audio
MPDU	MAC Protocol Data Units
MSC	Mobile Switching Center
mW	Milliwatt
NAT	Network Address Translation
NDEF	NFC Data Exchange Format
NFC	Near Field Communication
OBEX	Object Exchange Protocol
OFDM	Orthogonal Frequency Division Multiplexing
OpenSEA	Open Secure Edge Access
OSI	Open Systems Interconnection
P2P	Peer-to-Peer
PC	Personal Computer
PCI	Peripheral Component Interconnect
PCMCIA	Personal Computer Memory Card International Association
PDA	Personal Digital Assistant
PDCA	Plan-Do-Check-Act
PHY	Physical Layer
PIM	Personal Information Manager
PIN	Personal Identification Number
PPPoE	Point-to-Point Protocol over Ethernet
QAM	Quadrature Amplitude Modulation
RADIUS	Remote Authentication Dial-In User Service
RC4	Rivest Cipher No. 4
RFCOMM	Radio Frequency Communication
RFID	Radio Frequency Identification
RIM	Research In Motion
RAM	Random Access Memory
RSN	Robust Secure Network
RTD	Record Type Definition

RTS	Request To Send
S/MIME	Secure/Multipurpose Internet Mail Extensions
SAP	SIM Access Profile
SCO	Synchronous Connection Oriented
SDMA	Spatial Division Multiple Access
SDP	Service Discovery Protocol
SIG	Special Interest Group
SIM	Subscriber Identity Module
SMS	Short Message Service
SPAM	Spiced Pork And Meat
SPIT	SPAM over Internet Telephony
SSID	Server Set Identifier
SSL	Secure Sockets Layer
TCP/IP	Transmission Control Protocol/Internet Protocol
TCS	Telephony Control Protocol Specification
TKIP	Temporal Key Integrity Protocol
TPC	Transmit Power Control
UMA	Unlicensed Mobile Access
UMTS	Universal Mobile Telecommunications System
USB	Universal Serial Bus
URI	Uniform Resource Identifier
VLR	Visitor Location Register
VoIP	Voice over IP
VPN	Virtual Private Network
WAP	Wireless Application Protocol
WECA	Wireless Ethernet Compatibility Alliance
WEA	Wireless Application Environment
WEP	Wired Equivalent Privacy
Wi-Fi	Wireless Fidelity
WiMAX	Worldwide Interoperability for Microwave Access
WLAN	Wireless Local Area Network
WMAN	Wireless Metropolitan Area Networks
WPAN	Wireless Personal Area Network
WPA	Wi-Fi Protected Access
WPS	Wireless Provisioning Service
WSG	World Geodetic System
XOR	eXclusive OR

1

Introduction

◇◇

This book, in a condensed but comprehensive form, depicts the actual state-of-the-art technologies for wireless communication, with special emphasis on security aspects. Following are the subjects that have been covered in the book:

- Networks in General
- WLAN
- Mobile Phones
- Bluetooth
- Infrared
- Near Field Communication
- Emergency Management

The following fields have not been included:

- Radiation Scattering
- VoIP in Detail
- Skype.

WLAN being the basic and also most important subject, has been covered extensively, and, therefore, the chapter on WLAN is longer than those on other technologies. If one is interested only in a specific technology, it is not important to read the entire book but only that particular chapter, as each chapter is stand alone by itself.

Each chapter is inclusive of an introduction to technological principles, followed by possible risk scenarios, and organisational and technical countermeasures. Both risk scenarios and countermeasures, occasionally overlap between different subject areas or technologies. Since the book is structured along the lines of technologies and not security aspects, redundancies are inevitable. This is so, as each chapter is supposed to speak for itself.

It is also the case for comprehensive checklists appended to some of the chapters. They all start with strategic approaches followed by more technical details. The lists are presented in two column tables. The left column consists of questions, whereas the right one contains explanations (why the question is important?). If the question is relevant to security aspects, such as on serious threats, the next line in the table will contain a note in italics with the character of a warning.

Organisational measures have been proposed for many security issues. Therefore, in some places directives have been referred to. The chapter on mobile phones contains a simple directive, such as implementation rules. The chapter on comprehensive security policy structure emphasizes it to be imbedded into the general strategic corporate documentation. The introductory passages of the policy can be adopted more or less as presented. For details regarding a specific technology, a structure is provided, which can be completed by feeding in content from the preceding chapters.

Although many examples and scenarios have been taken from everyday operations of companies, including solution proposals, the security problems described are equally relevant to the usage of wireless communications by private individuals. Most of the questions in the checklists apply both to a single home station and large computer networks in companies as well. This is also true for the relevant technical countermeasures.

As an example, for a particular type of device, BlackBerrys were introduced with their own dedicated security philosophy. Additionally the state of the art of mobile phones has been taken into account as far as it has entered the market. In view of the short lividness of technologies this can, however, be only a snapshot, which hopefully will maintain some relevance for some time.

2

Network Security

2.1 Introduction

In economic life, industrial production, services, and other social relationships networks have become common structural elements. They have witnessed growth and have become an integral part of our lives, but there has been parallel growth in new kinds of risks and dangers associated with them. These risks may comprise hacking of web pages, applications being used by administrators or client data records from financial systems.

There is no such thing as absolute network security, only until one cuts off any external connections and prohibits all users. Once an organisation opts for internal and external networks—and usually there is no way around this—the need for risk assessment arises. This results in the identification of:

- Potential danger scenarios
- Countermeasures
  - Prophylactic
  - Reactive.

The structured document resulting from this effort contains some kind of security strategy for an organisation. Such a security strategy has to take care of not only that, but also ensure prevention of:

- Unauthorized access from external
- Data theft
- Data corruption.

The strategy has to be designed in such a way that the main objective of information processing, that is, the maximum availability of IT instruments is not impeded, taking into account all measures demanded by emergency management. This is the balancing act to be achieved.

Quite often, it is considered that technical measures, firewalls, encryption, etc., are sufficient for risk reduction. However, along with these measures, it is also required that the person concerned from IT eliminates potential security loopholes.

Generally this requires technical investments in objects trustworthy to guarantee the security requested. And there is no doubt about the fact that all necessary technical prerequisites have to be exhausted to enhance security. But on the other hand, even this approach is insufficient, as there are concrete demands from the user communities behind the obvious technical requirements surpassing purely technical aspects. Ideally, a security strategy should emanate from a catalogue of demands from the user community with relevant technical solutions covering parts of the whole package.

One must not forget that IT only provides some sort of infrastructure, such as hardware and communication facilities, which is only useful once applications using them provide some advantage to the user departments in their daily business. This means that besides application programs themselves the availability of databases and the possibility of transferring information internally and externally are indispensable.

The procedure to develop a network strategy corresponds to the outline in Fig. 2.1.

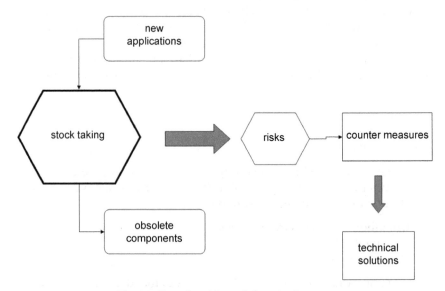

Fig. 2.1. Procedure Network Security Strategy.

First and foremost it is important to document the application landscape as a whole. Thereafter, possible threats need to be identified. This further leads to countermeasures. Only after these steps the technical solutions to be deployed are identified. In between all these efforts, it is important to involve the user community from the very start to protect their interests as well. This is best served by drafting appropriate specifications. How these demands will be implemented in the end are laid down in technical specifications by the technicians themselves, to be cleared in turn by the users again.

Furthermore, the development of such a concept offers the opportunity to include already future system and application enhancements. In this way the security strategy is closely linked to the overall IT strategy of an organisation. All these considerations imply that the drafting of such strategic papers means a lot of effort—efforts far surpassing the one-time installation of a new firewall. It is just as everywhere: the work has to be done. Things, which have been neglected during preparation, have to be retrofitted later at higher costs.

This means that during the development of a security concept even existing technologies have to be questioned for:

- Outdated applications
- Unsuitable hardware
- Unsuitable communication equipment, etc.

As regarding security relevant objects, they comprise:

- The whole area of a company/organisation
- All buildings; especially premises which permit direct or communicative access to computer systems and communication facilities
- Utilities
- All hardware in connection with information and communication, movable or fixed
- Closed surroundings of the company compound, in as much as wireless access may be attempted to internal systems.

It becomes quite clear that a network security strategy does not primarily require a purely technical concept, but is first of all an organisational challenge. Figure 2.2 shows the context between the different strategic levels.

When establishing objects to be protected a simple identification does not suffice. The resources to be protected may be subject to different endangerments. In contrast to classical networks relying on their own proper wiring and housed closely within the premises of an organisation, today endangerments have erratically risen in many cases due to Internet linking

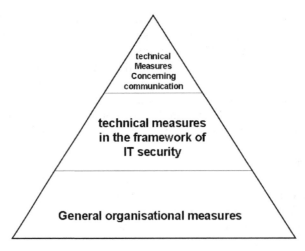

Fig. 2.2. Security Pyramid.

and use of mobile terminals. Every mobile terminal (cell phones, smart phones, tablets, etc.) is not only equipped to call but is fitted with capabilities to permit its use as a fully functional terminal to central applications. In this way endangerments and motives differ, and thus the types of attack as well.

2.2 Weaknesses in Network Security

Networks are part of the business world, including the remote managing of branch offices, controlling mobile workforces, and home offices. But all these developments also entail new risks, with attempts of eavesdropping along communication lines and attempts to gain unauthorised access with remote methods, thus endangering databases and the networks of companies.

The problem is that network access to data and programs is required to be provided at any time and anywhere. The transmission of data and other information is part of working procedures within all areas of an organisation. In fact the functioning of a company nowadays is strongly dependent on network transmissions and its associated hardware. In case of disruptions the incurring costs are mostly unknown but certainly not insignificant. All this necessitates the implementation of highly professional security arrangements in networks. These provisions should not only protect against disruptions or complete breakdown but also ease the complex tasks of network management, and prevent corruption and theft of data through unauthorised access, thus furthering the common interests of a functioning organisation.

Tackling the above cited challenges with technical means is common practise in most of the companies. Implementation is delegated to the technical departments. When there are problems concerning computers or databases in networks, it is managed by the IT department. By their very nature these resources concern themselves primarily with the characteristics of technical components including not only the complete hardware and software environment, but also service arrangements and backup strategies. This is approach is called "bottom up". The general approach to security problems starts invariably with the improvement of firewalls.

Of course technical approaches to security problems are indispensable and form the backbone of any security strategy, but they should not be the only base for general considerations and decision making. On top of that other aspects, besides the purely technical ones, also play an important role. In fact it should be the other way round: technical solutions come into play only after all other frame conditions have been taken into account. In the first place a company should become aware of security arrangements necessary with a view on its business. These considerations should not be driven by technical possibilities alone. How the proposed requirements could be satisfied by technical solutions is a consequence of the strategy as such.

Operational procedures in the daily business of users are unthinkable without technical support. On the one hand there is a whole hardware infrastructure inclusive of:

- Cables
- Network components
- Computers with their operating environment and protocols.

All these constitute the basis of a network and in a sense support the daily life of the users. But the real, visible elements concerning procedural support are:

- Applications and
- Databases.

The user relies on the availability of databases and, for instance, also support through his mailing system.

This implies that any security concept has to start with the application level. Otherwise the objective of problem-free staff support cannot be implemented. From demands put to applications and service requirements relevant to a network can be deduced in turn. This also means that quality and security of the application environment are closely connected. The planning process for this approach should at the same time include the anticipation of possible future problems. It has to be integrated into an overall strategic concept of the business. All this goes not without costs. Significant efforts have to be exerted for detailed analyses. These efforts pay off in the end, firstly because it eases the smooth implementation of the measures in question, and secondly it facilitates detection and correction of errors during later operation of the . system.

Of course, internal and external protective measures are an integral part of any such concept. The whole question of security encompasses more than just technical means. It includes all necessary managerial and organisational elements as well. This in turn demands the cooperation of all relevant company resources besides the technical departments. To begin with the company leadership or a delegated project management has to find out about the requirements of staff at all levels. Therefore, all relevant daily routines have to be recorded. However, it remains for the IT department to later budget all necessary protective technical means to serve these requirements on the network level itself. Last but not the least, the analysis should include possible alternatives for outdated applications and systems. During the process of analysis the opportunity to evaluate outdated applications should not be left out.

2.3 Relevant Resources

There are two different venues to attack a system: from outside (mostly discussed in the public), and also from inside sources. Listed below are the most important resources:

- Hardware
  - Workstations
  - PCs
  - Other terminals
  - Servers
  - Peripheral equipment
  - External storage media
  - Communication lines
  - Switches
  - Routers.

- Software
  - Programs
  - Utilities
  - Operating systems
  - Communication protocols

- Data
  - Emails
  - Backups
  - Audit logs
  - Databases

- Staff
- Security relevant documents.

2.3.1 Danger Potentials

There are certain danger potentials attached to the resources identified above. Danger potential refers to a possible loss of the resource in question. They include attacks that do not alter anything in the system, such as:

- Eavesdropping of
  - Any information concerning the business of a company
  - System parameters
  - Login passwords
  - Private data.

This information can be obtained by employing network analysing tools or fake identities.

Other attacks may result in

- Modification
- Deletion
- Insertion

of data or programs.

Finally there is a third category concerning danger potentials: the accidental falsification, which may take place unintentionally, for example:

- Erroneous routing
- Transmission of wrong data
- Program errors
- System errors.

In the end error rates and attack probability rise with the size of an installation and the bulk of data to be transmitted.

2.4 Objectives of a Security Strategy

When developing a security strategy of any kind the objectives for the company have to be formulated in an unambiguous way. They should take into account the following aspects:

- **Availability**

  As has been pointed out above, risk potentials rise with the number of services provided for. Thus, any service benefit has to be weighed against its associated risks. If it becomes evident that security costs surpass the benefit of a service, the deactivation of the latter or the implementation of a less costly alternative have to be considered.

- **Access and security**

  The most user friendly system would relinquish any access control by passwords, etc., and grant access to anyone. Unfortunately, only login procedures, as we know them, grant some limited security against unauthorised access. However, without frequently changing passwords, even this type of simple security may become a loophole. Resources are just not freely available; full stop.

- **Security vs. risk**

  To protect a complete system various elements have to play together, all of which generate costs. These costs include:

- Costs for the acquisition of hardware
- Software for firewall building
- Password generator.

A secure encryption of data results in a lower throughput and asks for hardware and software enabling better performance. Furthermore, data encryption usually entails performance degradation. This also has to be taken into account while decision making. Risks to data include:

- Loss of privacy
- Data loss as such
- Loss of other resources by alterations of system parameters.

2.5 Security Aspects Concerning the Internet

Some parts of our society do no longer function without the Internet. The Internet not only furnishes news and entertainment, but is also the basis for vast numbers of business transactions. This starts with the presentation of a company on its homepage. Real security aspects become important once, for example, customer or supplier data is exchanged. Thus, the Internet and its mechanisms may serve as a gateway for illegal access to company information. The task is to identify these dangers and counteract them.

The most common types of dangers are represented by viruses and associates. Organisational and firewall mechanisms may impede this danger. But there is a possibility of other serious types of attacks via the internet:

- Entering central applications via the Internet and misusing them
- Downloading a malware-program from the Internet.

Attacks via the Internet may enter a company's internal network as well. There are specific programs available to carry out such attacks or, of course, by manual endeavour. Any operating system or transmission protocol contains weak spots and loopholes through which a determined and well-informed attacker may try to gain illegal access. This includes users, who, by chance, get acquainted with hacking software, and are tempted to try out something. It may result in essentially harmless intrusions, but it can also cause the same amount of work for security officers to undo the damage done by serious attacks. A professional hacker, however, will try everything to extinguish any traces of his actions.

Besides immediate damages (loss of data, data theft, etc.) there may be other damages as a result of network break-ins: after an attack or a break-in system integrity is required to be restored, which can be quite costly. There will be directly visible results of a hacking attack in the form of data corruption or system failure. However, other questions remain:

- For how long did a hacker remain in the system?
- Which data have been manipulated?

To answer these questions, a lot of research effort has to be employed with all associated costs.

Again we are back to the requirement of a comprehensive security strategy.

The network operator has to consider carefully, which resource would be the target of an intruder. Internet risks lead to the same questions as otherwise:

- Which type of data could be targeted?
- What would be the consequences, if these data are altered?
- Are personal data concerned?
- Do security levels already exist in the company?

These questions have to be answered in detail to safeguard the resources in question, and their answers have to be integrated into the security concept. All these aspects cannot be generalised. Their importance and the answers to them may vary across different companies. Here are some of the key aspects concerned:

- Confidentiality
- Integrity
- Availability.

As pointed out already, the first and foremost measures to be taken can be found in the following three basic requirements:

- Identification (user identity, including password)
- Authentification (verification of user data entered) by
  - Information entered by typing
  - Special chip cards
  - Dongle
  - Biometric data.
- Authorisation to use those resources allocated to that particular user.

To achieve the above, access control systems are employed. These systems contain all necessary information to verify access security. They include the following components:

- A database with access rights mapped to a user identity
- A surveillance and authorisation system using the database following specific rules laid down in connection with users; it also monitors updates of user information.

In the end hard decisions have to be taken regarding the key aspects bound to any resource:

For example, does confidentiality have a higher priority as availability or vice-versa? This question has to be answered by security officers once a break-in has been detected to decide on the need for a shut down. The answer to this dilemma is not always straightforward, but depends on the degree of encryption of information and on the necessity to continue commercial activities. Depending on the type of installation the relevant decision making rules are part of the security documentation and also serve as a basis for the selection of technical tools to be installed.

We have already mentioned the types of intentional and unintentional unauthorised access to data and programs. Here some of the specific methods to be observed:

- Man in the middle attack.

The attacker connects himself between two communication partners and deceives both of them to be the real partner in question.

- Deception.

An attacker inserts messages into a data flow or deletes or modifies legitimate messages.

- Replay

A legitimate message is repeated later by modifying the system timer or job control statements.

- Password guessing
- Cryptographic analysis by statistical means without knowing the encryption key.

To summarise everything said so far in a very general way (more specific details as to types and methods of attack and countermeasures will be presented in the chapters dealing with the various communication protocols), these are the salient points around which any security strategy should be centred in a network:

- Protection against external hackers
- Protection against unauthorized manipulations from within
- Protection against viruses
- Backup routines.

3

WLAN

◇◇

3.1 WLAN Basic Principles

Interconnections of computers and their components has reached a new level of quality for private users and organizations as well with the deployment of wireless technologies. The development of the WLAN (Wireless Local Area Network) was a milestone in this process. WLANs bring with them their very specific security requirements, which is the subject of this section of the book.

The following aspects will be covered in detail:

- WLAN general features
- Security requirements
- Overview of the relevant standards
- ISO layers and encryption.

Supplemented by:

- WLAN architectures
- Components
- Configuring a WLAN
- Application examples, and
- A security checklist.

3.1.1 Advantages of Wireless Networks

Making cables obsolete not only saves investing in them, it also offers mobility of a kind previously not experienced by the user. The typical image of a person sitting in the garden at home at work with a notebook illustrates this convincingly. Imagination carries this even further by suggesting this possibility—via notebook or smartphone – in any place in the world near a hotspot to link up to the company network or the Internet.

3.1.1.1 Mobility and Portability

Besides the changes to working processes already triggered by mobile phones, mobile networks offer an additional push to develop business processes further. This is true, for example, in the case of large building sites, for managing sizeable stores, and also for the functioning of medical centers or clinics. At times, quick access to medical data can be life saving too. Through mobile link-ups work interruptions can be minimized, since the required information is accessible from any given location at any time.

Mobile types of networks permit access to data in real time—just as with a classical LAN. But mobile networks gain importance, where, for example, hard wirings require high technical expenditure or special provisions at protected architectural sites. In any case the WLAN solution makes even more sense when setup and operations are only of a temporary kind, such as at trade fairs or for project teams working together only for a limited period of time. Not only can the cabling be saved, but radio networks can be set up much quicker than classical ones.

WLAN applications have experienced a boom in the private sphere. But since quite often professional experience is lacking there, the security risks are higher.

3.1.1.2 Security

A wireless connection is exposed to other endangerments than fixed networks. The reason lies in the choice of radio itself. An attacker could eavesdrop on an application in someone's home from a nearby car park by using his notebook and a chips can with an antenna.

Such activities are called "wardriving". Meanwhile the Internet offers pages listing unprotected WLANs in certain towns and regions. Therefore, there is some urgency to secure radio networking by developing appropriate protective measures.

Wireless access to data in local networks or via the Internet is not critical in terms of security, if the information in question is publicly available in the first place anyhow. However, regarding the exchange of sensitive data, there are additional security challenges with regard to ordinary computer networks. These are due to the specific dangers brought on by the technology itself. Radio waves as carriers of information can be tapped and disrupted.

The security provisions offered by the first generations of WLANs soon showed major flaws in practice. The encryption protocols known under WEP offer weak protection and can be broken relatively easily.

Only in 2004 did the IEEE publish specification 802.11i with a standardized security architecture as a solid basis for reliable WLAN solutions. But even today not all WLAN components in the market do adhere to this standard. Therefore, intermediate solutions, such as WPA play a continuing role.

3.1.2 General Features of Wireless Networks

A radio network offering functionality similar to a LAN is called a Wireless Local Area Network, wireless LAN or WLAN [1]. The naming of the network as "wireless" is somewhat broader than the term "radio net", since this may also include infrared.

In practice radio nets are usually coupled to wire bound networks thus complementing the LAN to allow for more mobility for certain users. In case of several LANs coupled together they are also called MANs (Metropolitan Area Networks) [2].

3.1.2.1 Radio Frequencies Available

The spectrum of electromagnetic waves used for communications can be differentiated according to frequency and wavelength. For radio and television broadcasting frequencies between 30 kHz and 300 MHz (long, short, and ultra short wavelengths) are used, whereas wireless networks work with shorter waves and thus higher frequencies between 300 MHz and 5 GHz.

In 1985, the Federal Communications Commission (FCC) released the ISM (Industrial, Scientific and Medical) band for general use in North America [3]. Frequencies within this bandwidth can be used free of license. Concerning WLANs the frequency range is between 2.4 and 5 GHz. This decision by the FCC opened the possibility for industry to develop inexpensive components for WLANs.

3.1.3 Standards [5,6,7,8]

Communication within a computer network—cabled or wireless—is impossible unless certain rules are followed. These rules are defined in protocols, which are acknowledged globally. For WLANs most of the LAN protocols are relevant. In addition, protocols concerning the special features of wireless networks have become necessary. To understand how to configure and run them securely a certain understanding of these standards is required.

Only at the end of the eighties of the last century did the IEEE begin to develop suitable standards. In 1997, the first WLAN standard was published under the number 802.11 [4]. In the years that followed several enhancements and extensions were made public, identified by a special letter at the end of the standard number.

In 1999, the IEEE published two more standards, the 802.11a using frequencies within the 5 GHz range, and the 802.11b. The latter is the standard most widely in use today. This goes for private applications, companies and publicly available hotspots as well. 802.11b allows for gross transmission rates of up to 11 MBits/s, most of which is used up for protocol overheads. 802.11b works within the 2.4 GHz frequency range and uses the HR/DSSS procedure. In 2003, the standard 802.11g was adopted, permitting transmission rates of up to 54 MBits/s within the same frequency range. In 2004, the standard 802.11i offered improved safety architecture. The 802.11n operates again in the frequency range of 2.4 und 5 GHz, but has a higher transmission range (600 Mbit/s) and a reach of up to 300 m. In 2010, the 802.11p was released to operate in vehicles. To suit the requirements of smartphones, more recent developments resulted in the 802.11ac and 802.11ad. Further details to these standards will follow later in this section.

3.1.3.1 ISO and 802.11

The standards of the 802.11 group follow the ISO definitions (Open Systems Interconnection Reference Model) by the ISO (International Organization for Standardization) [9]. This rather abstract model describes the communication between open and distributed systems on a functional basis along seven layers of protocol built upon each other. Open means that the model is not bound to certain company standards, distributed means a decentralized system environment.

3.1.3.2 PHY

If an application attempts to start a communication between two addresses within a network all the ISO layers—each with its special assignment—are passed through in sequence. The lowest layer is the physical layer (PHY). The protocols belonging to this layer describe connection set-up and connection tear-down between the components in question and the transposition of data into physical signals, them being electrical, electromagnetic, or optical.

3.1.3.3 Connection Control

Above the physical layer resides the Link Layer called Data Link, responsible for handling the connection between the sending and the receiving entity and for the reliability of data transmissions. The relevant protocols are used to organize the transformation of the data into packets. On top of this the transmission of those packets are monitored using functions that are able to detect and sometimes even correct transmission errors.

The third layer, the network layer, takes over the routing process for all error-free data packets. It is followed by the protocols concerning transport, session, presentation and application, all of which will not be discussed here any further.

3.1.3.4 MAC [10]

WLAN specific standards only deal with the two bottom layers. As can be seen from Fig. 3.1, the 802 family of standards only deals with the physical layer and part of the link layer. For this purpose, the link layer is subdivided into two sub-layers: MAC (Medium Access Control) and LLC (Logical Link Control). LLC is defined in a separate standard 802.2 [11] for all types of local networks. Its protocol manages the connection between sending and receiving computers.

MAC ensures data packaging into so called MAC protocol data units (MPDU) [12] and controls access to the carrier medium and the mode of operation, which has been defined in the physical layer. The appropriate rules come into play, when several stations share the same carrier medium. The purpose of the MAC protocols is to avoid collisions and in consequence data loss. This could happen, if several stations within the same network try to send and receive data at the same time. To make sure that for a given point in time only one single device is sending, the standard 802.11 uses the CSMA/CA (Carrier Sense Multiple Access with Collision Avoidance) [13] procedure.

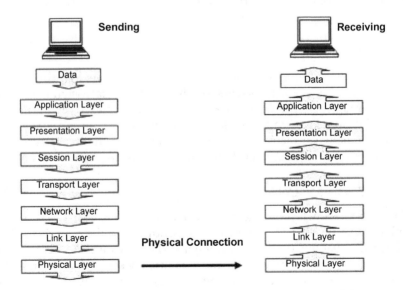

Fig. 3.1. Data Transmission According to the ISO Reference Model.

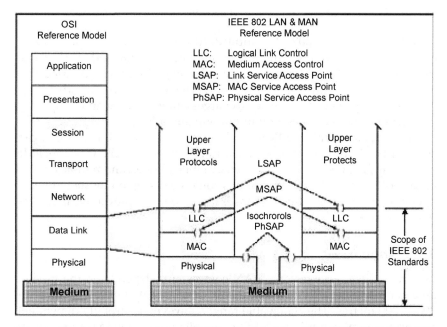

Fig. 3.2. Relationship between the ISO Reference Model and the IEEE 802 LAN/MAN Reference Model (IEEE Std 802-2001).

3.1.3.5 WLAN as Part of LAN

The advantage of the ISO Reference Model lies in the fact that protocols of the upper layers can access services of the respective layer below and, thus, do not have to bother about these tasks themselves (Fig. 3.2). A network link can be assigned to the link layer to link LANs with different physical characteristics. By contrast routers are assigned to the network layer.

3.1.4 Connecting to Computers

The IEEE 802.11 specification belongs to the LAN/MAN standards all grouped under 802. Initially the standard dealt with components of wireless networks allowing for transmission rates between 1 MBit/s and 2 MBit/s. The envisaged radio technologies included Frequency Hopping Spread Spectrum (FHSS) [14] and Direct Spread Spectrum (DSSS) [15] as frequency spreading alternatives.

The later versions of 802.11 specified wireless connections with transmission rates of up to 11 resp. 54 MBits/s using different techniques of frequency modulation: High Rate Direct Sequence Spread Spectrum (HR/DSSS) and Orthogonal Frequency Division Multiplexing (OFDM) [16].

3.1.4.1 802.11 and ISM

By adhering to the 802.11 standards the interconnect capability of components from different manufacturers can be guaranteed. At the same time the use of the ISM band has the advantage that it can be used free of license in most countries of the world. This is an enormous advantage especially for private users since no charges come into play, and temporary networks can be put together without bureaucratic barriers.

The problem with the IMS band, however, lies in the fact that it can be used by a number of other technical devices as well. These include medical appliances, microwave ovens, mobile phones, and remote control of car ports. Thus, WLAN operations could be disrupted by such devices operating close to the frequency ranges in question. So, preventive measures have to be taken.

3.1.5 Antennas

The transmission power of WLAN components depends on the type of antenna employed. The state-of-the-art components fall in the range of 200 to 300 metres. The achievable range, however, is strongly dependent on various local conditions. Additional antennas enable to bridge longer distances. Directional radio antennas can cover several miles. The increase in range is called antenna gain.

3.1.5.1 Antenna Types and Antenna Gain

Antenna types are classified according to propagation patterns and amplification method. One distinguishes between omnidirectional and unidirectional antennas. The latter have a much longer range, since they operate only at a narrow aperture angle with the same transmission power. This is another form of antenna gain.

Access points usually rely on omnidirectional antennas. Directional antennas are only used to overcome longer distances. Instructions to build antennas can be found on the Internet. This poses an additional security problem.

3.2 IEEE 802.11 Overview

The standards for wireless networks, which were published by the IEEE in recent years, were—as already mentioned—from the beginning part of the 802-family to facilitate the connection to classical Ethernet solutions.

WLAN components on the market are able to deal with different versions of the 802.11. Independent of the compatibility question they all have in common that on the type label or at any other suitable place that version of the standard is marked, which can be served by a particular device: 802.11b for example. 802.11 means the standard itself; the suffixed letter stands for a particular version. It was originally derived from a specific task force that developed this particular version within the framework of the overall work group responsible for all WLAN standards. These versions shall be explained in detail in the following sections.

The version 802.11b is actually the one disseminated most. This is due to the frequency band. The 2.4 GHz band is available licence free in most countries. There are of course new components supporting other members of the 802.11 family (for example, 802.11g, 802.11i). Some of the versions are compatible between them. Furthermore, one can purchase components, which support different versions at the same time. There are access points for 802.11a, b, and g.

Today there are WLAN interfaces in all notebooks and smartphones in the market.

Table 3.1 gives an overview about the historical development of the IEEE standards in question:

The standards of the 802.11 family have been tailored to make wireless communication compatible with Ethernet solutions.

The 1997 Version

In 1997, the first version of a WLAN standard was released under 802.11. The document outlines the data handling in the wireless context with respect to the ISO physical layer. The specification details the frequency spreading methods FHSS and DSSS. Originally, the standard proposes gross transmission rates of 1 MBit/s for FHSS and 1–2 MBit/s for DSSS.

Communication between any two participants can function either in the so called ad hoc mode or via base stations (access points). The latter is called infrastructure mode.

When buying a WLAN component, the device designation is preceded by a reference to the standard version, for which the device has been released, for example: 802.11b Wireless USB Adapter. The figures relate to the IEEE standard for WLANs, the letter indicates one of the various versions of the standard. These letters again refer to the different Task Forces within the IEEE Work Group for WLAN Standards. Further down the different versions of standard development since 1997 will be outlined.

Since in most countries the 2.5 GHz band can be used license free and without restrictions, chiefly products using the standard 802.11b are in circulation. More recent products support 802.11g or even 802.11i. This raises the question of compatibility between the various versions. More and more products have entered the market supporting several different versions. Netgear, D-Link, Lancom, or 3Com offer access points for professional use supporting operational modes for 802.11a, b and g or 802.11b and g or alternatively b or g.

Table 3.1. Development of 802.11 Versions.

Standard	Year	Features	Frequency Range GHz	Transmission Rate MBit/s
802.11	1997	First version, physical layer, FHSS, DSSS	2.4–2.485	1–2
802.11	1999	Revision	2.4–2.485	1–2
802.11a	1999	OFDM	2.4–2.485	6, 9, 12, 18, 24, 48, 54
802.11b	1999	HR/DSSS	2.4–2.485	5.5, 11
802.11d	1999	MAC, international harmonization	2.4–2.485	5.5, 11
802.11g	2003	OFDM	2.4–2.485	6, 9, 12, 18, 24, 48, 54
802.11h	2003	MAC, TCP, DFS	5	54
802.11i	2004	WPA	5	54
802.11n	2010	MIMO, channel bonding	2.4, 5	100
802.11p	2010	Communications in vehicles	5.850–5.925	27
802.11ac	2013	Extension of 801.11n	5	6936 max.
802.11ad	2014	Large Bandwidth	60	6930 max.
802.11ah	2016	Smart Homes	0.9	0.65–7.8

Some notebooks are equipped with integrated WLAN functionality by Intel's Centrino Technology [17], so that a special adapter is not needed. Besides the support for 802.11b, a two band solution 80211.a/b is possible supporting both modes 802.11b/g.

3.2.1 The Standard 802.11 and its Extensions

Note: here the performance characteristics of the different versions of the standard and their improvements are outlined successively, while going through these versions. The respective security relevant features are detailed later in Chapter 5.

The specifications of 802.11 define, how to set up, maintain and tear down a connection within a radio network for the physical layer. For the link layer only the medium access control to the WLAN medium has been defined.

The second partial layer of the ISO link layer (LLC) is not specifically addressed by 802.11 (Fig. 3.3). For this the 802.2 standard initially developed for LANs has been adopted. In other words, the MAC layer access protocols have to deliver results compatible to a LAN:

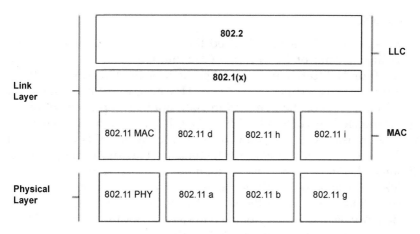

Fig. 3.3. Standard Versions in ISO Layers.

3.2.1.1 802.11

The original Standard 802.11 of 1997 defined the modulation methods FHSS and DSSS for radio networks with respect to the physical layer. Data transmission rates are limited to gross 1 or 2 MBit/s. The transmission band is 2.4 to 2.485 GHz.

The standard also specifies the operational modes "ad hoc" and "infrastructure". For the MAC layer access rules are defined and the WEP encryption procedure [18] is offered as a possible security mechanism. This basic version of the standard was revised in 1999 superseding the text of 1997. Components corresponding to these basic specifications are rarely being used today.

3.2.1.2 802.11a

Already in 1999 the basic specification of 802.11 was enhanced by two new versions: 802.11a and 802.11b. With 802.11a, the OFDM procedure was added to the physical layer as a new feature. OFDM allows a much higher data transmission rate and uses a different frequency band. Gross data rates available from then on are: 6, 9, 12, 18, 24, 48, and 54 MBit/s (proprietary extensions by some manufacturers even obtain 108 MBit/s). The stations always try to operate at the maximum possible rate. If they are located close to each other a higher rate is used. If transmission errors occur because of range problems, the transmission rate is reduced.

3.2.1.3 802.11b

The version with letter b, published the same year, still works within the frequency band of 2.4 to 2.4835 GHz, but improves the data rate to about 5.5 or 11 MBit/s gross with the help of the HR/DSSS method, being downwards compatible to DSSS. However, this only yields a net rate of about 50% of the nominal. In 2001, another improvement of the specification followed. With 11 MBit/s the WLAN standard attained about the same transmission rate still widely used today in LANs with around 10 MBit/s. The rate is in any case faster than most Internet connections but still somewhat too slow to transmit audio-visual media data.

802.11b is actually still the standard, which is used most widely. This can be attributed in part to the WECA (later called Wi-Fi Alliance) [19], founded in 1999, which has promoted the technology by issuing the Wi-Fi logo. Although the transmission rate (11 MBit/s) is considerably less than with 802.11a this standard has the advantage of a longer range between different buildings or outside. Since the 2.4 GHz frequency band is also used by devices working with Bluetooth or mobile phones interferences are possible.

3.2.1.4 802.11d

This specification adds rules to the MAC layer to enable the world wide deployment of WLAN components. Adaptations of sending parameters and roaming facilities have been added.

3.2.1.5 802.11g

In June 2003, the IEEE released a new WLAN Standard being downwards compatible with 802.11b and again operating in the frequency band 2.4 to 2.4835 GHz. It uses the OFDM method and allows the same transmission rate as 802.11a with a maximum of 54 MBit/s. The range, however, corresponds to that of the 802.11b. The new 802.11g components, thus, fit easily in existing WLANs made up of 802.11b components. However, an 802.11g device then changes into a special compatibility mode, meaning that the effective transmission rate is reduced to 10–15 MBit/s.

3.2.1.6 802.11h

With the specification 802.11h some adjustments have been made to the MAC layer for WLANs in the 5 GHz frequency range to take into account European radio regulations. This standard uses the Transit Power Control Procedure (TCP) to reduce the transmission power as a function of connection quality similar to mobile phones. This procedure had been demanded by ETSI for WLANs under 802.11 [20].

TCP makes sure that the required transmission power between communicating stations or between access points and a station remains within a permitted range. To dynamically adjust the transmission power the stations can request information about the connection route between corresponding stations via special TCP Request Frames.

Dynamic Frequency Selection (DFS) is a technique selecting the best frequency at any given time. The DFS automatically changes the channel, if during the utilization of a specific channel another user or technical device is detected (for example, a radar unit) also operating within 5 GHz frequency band. Before occupying a channel a test is made to find out, whether a different system uses the same frequency range. In this way unwanted interference of the WLAN within the 5 GHz can be avoided. 802.11a and 802.11h are otherwise compatible between one another.

3.2.1.7 802.11i

IEEE 802.11i has a new security protocol released in June 2004. It has been developed as an alternative for the encryption procedure WEP, which has been criticized heavily. WEP is not simply substituted by an alternative encryption procedure but by a complex security architecture as its name indicates: Robust Security Network (RSN). This protocol is relevant for the standards 802.11a/b/g/h. Providers like Lancom and others offer the necessary firmware.

802.11i makes it possible to protect even the ad hoc mode. Besides other encryption procedures it uses the Advanced Encryption Standard (AES) [21] and offers cipher key management using TKIP and a secure encryption method for WLAN access defined by the general network standard IEEE 802.1x based on the Extensive Authentication Protocol (EAP) [22]. This standard is not part of the 802.11 family but belongs to the superior 802 family relevant for networks in general. It describes possible procedures for the authentication and authorization of users and components in local networks mainly based on RADIUS [23] and EAP.

RADIUS (for Remote Authentication Dial-In User Service) is a de facto standard for authentication systems with dial-in connections, requesting user name and password for network access. This information is handed over to a RADIUS server for verification and approval. Communication between RADIUS client and RADIUS server is carried out in encrypted fashion. User data are thus not transmitted in plain text as is done frequently in other log in procedures.

EAP is a general protocol for the authentication of users in a network supporting various authentication methods. It is an extension of PPP [24]. If a WLAN user tries to establish a connection to an access point the user is requested to identify himself, and the corresponding information is transmitted to an authentication server.

Part of 802.11i had been anticipated by the Wi-Fi Alliance under WPA. Therefore the Alliance talks about the new standard as WPA2 [25].

3.2.1.8 802.11-2007

In this version of March 8, 2007, eight extensions (802.11a, b, d, e, g, h, i, j) were consolidated.

3.2.1.9 802.11n

As all other standards the 802.11n operates in the frequency ranges 2.4 and 5.0 GHz. The objective of this new development was a transmission rate of 600 MBit/s and a reach of 300m. These are, however, theoretical values. In practise, the rate is around 100 MBits/s. This is due to the interplay of different components in networks. As this standard is backwards compatible to 802.11a, b, and g, the rate is further reduced.

The standard employs basically three technologies: Multiple Input Multiple Output (MIMO), Channel Bonding, and Frame Aggregation. MIMO uses several transmitters and several receivers at the same time. By space multiplexing the data streams are fragmented and transmitted as individual units via the same channel. From these streams the receiver consolidates the message by using a complex algorithm. Furthermore, MIMO focuses the energy of the radio signal in the direction of the receiver in question. The Channel Bonding method of the 802.11n combines two 20 MHz channels to a single 40 MHz channel and, thus, doubles the transmission rate. The transmitted volume can be augmented by combining individual frames to larger data packages and thus reducing the overall number of frames and also their overheads.

3.2.1.10 802.11p

This standard is an enhancement of 802.11a to be used in vehicles to permit communication between different vehicles, and was released in 2010. The data rate is 27 MBit/s brute within the frequency range between 5.850 and 5.925.

3.2.1.11 802.11-12

This standard consolidates 10 extensions to the 802.11-2007 (802.11k, r, y, n, w, p, z, v, u, s) and was published in March 29, 2012.

3.2.1.12 802.11ac

This is an extension to 802.11n, released in 2013. The data rates are 6.5 to 96.3 MBit/s by 20 MHz channel width, 13.5 to 200 MBit/s by 40 MHz channel width, 29.2 to 433 Mbit/s by 80 MHz channel width, 58.5 to 867 MBit/s by 80 MHz or 160 MHz channel width and up to 1299 (theoretically 6936) MBit/s for devices equipped with MIMO and a channel width of 80 MHz.

First devices (router, laptops, smartphones) entered the market at the end of 2013 using the frequency range of 5 GHz.

3.2.1.13 802.11ad

This version uses a larger bandwidth with four channels within the 60 GHz band. The data rates for OFDM mode are 1540, 2310, 2695, 3080, 4620, 5390, and 6930; for QAM mode: 26, 361 and up to 5280 MBits/s with a maximum reach of 10 m.

3.2.1.14 802.11ah

This standard was released in August 2016. It operates in the frequency band of 900 MHz with 26 1 MHz channels resp. 13 2 MHz channels.

3.2.1.15 Non Standard Versions

Some manufacturers of components and solutions in a specific field think that the functioning of large standardizing committees like IEEE are too slow. This is the reason why sometimes proprietary extensions are on offer. This is also true for WLANs. For example, a version with the designation "802.11b+" has been developed to permit higher transmission rates (from 22 up to 44 MBit/s within the 2.4 GHz band). There are versions for 802.11g with the bundling of two channels and a rate of up to 108 MBit/s. All these are non-official standards concerning the IEEE. And there are possible consequences regarding the compatibility with components of other manufacturers.

3.2.1.16 Alternative Standards

Besides the standardizing projects of the IEEE there have been other efforts to formulate rules for close range wireless data transmission. This was attempted in the first place by the European Telecommunication Standardization Institution (ETSI) having published HIPERLAN/1 [29] already in 1998 and the HIPERLAN/2 specifications in 2000. HIPERLAN stands for High Performance Radio Local Network. The Home Radio

Frequency Group published HomeRF [30] for transmission rates of 10 MBit/s in 1998. However, all these alternatives had no significant impact.

3.3 Wireless Fidelity

In 1999, the Wireless Ethernet Compatibility Alliance (WECA) was founded by companies active in the WLAN market. They created the label Wi-Fi (Wireless Fidelity). Later the Alliance was renamed Wi-Fi Alliance. This Grouping wants to make sure that WLAN components correspond to the 802.11 standards.

Such certified WLAN components from different manufacturers can thus be combined as long as they are operated within the same frequency band.

3.3.1 Wi-Fi Protected Access

Because the weaknesses of the 802.11b WEP encryption became known soon, the Wi-Fi Alliance had introduced an alternative under Wi-Fi Protected Access (WPA) even before the IEEE itself could release an improved standard. A corresponding certificate was issued for devices adhering to this alternative: Wi-Fi CERTIFIED for WPA. WPA anticipates part of what was later contained in 802.11i in 2004. After 802.11i the Wi-Fi Alliance offers a new certificate: Wi-Fi CERTIFIED for WPA2.

3.4 WMAN

The next generation of WLAN development comprises concepts for radio networks, which can cover longer distances. A special IEEE task force has been created (Task Force 802.16) to deal with Wireless Metropolitan Area Networks (WMAN). These networks will cover areas with a radius of about 50 km with a transmission rate of 70 MBit/s. The frequency range is between 10 and 66 GHz. In 2002, a first draft was issued, followed by 802.16 in 2003. The latter document included frequencies between 2 and 11 GHz. This is the same range used by WLANs. Components on the market today are at present not a serious threat to DSL.

Similar to WLANs with its Wi-Fi Alliance there exists a corresponding interest group to further the proliferation of WMAN—the WIMAX Forum for Worldwide Interoperability for Microwave Access [31].

3.5 Key Terminology

3.5.1 WLAN Components

To build a WLAN one needs certain components, some of which will be described in detail later. These components are required to provide an organization or parts of it with a wireless network. On the other hand, they could in turn be used to connect mobile terminals in certain locations to central applications. The utilization of a WLAN could be offered to third parties commercially. Ad hoc connections between any two or more partners can be established for example via integrated WLAN capabilities into smartphones.

As a minimum one needs appropriate network adapters installed on mobile devices. For notebooks not already equipped with integrated WLAN capabilities one generally uses USB adapters or PCMCIA cards. The associated drivers assure the necessary send/receive functionality. Most recent notebooks have WLAN components already integrated.

3.5.1.1 Access Points

In case the mobile devices do not only want to exchange data between themselves, stationary components are required to serve as an interface to a LAN. These are called access points (AP). Quite often these components are integrated into a more general device offering router functions, hubs, DHCP server functionality or a DSL modem. These devices also support the NAT function (Network Address Translation) [32], i.e., for example, the possibility to work with different IP addresses within the same network, but externally showing only a single IP address to the Internet. This not only permits collective access to the Internet, but can also function as a firewall to prevent undesirable access from the Internet to the various stations of the network.

Access point functionality can also be realised by special software on a PC.

3.5.2 Bandwidth

An important role is played by the carrier medium transmission capacity. Any possible data rate depends on the bandwidth, i.e., the frequency range of the signal transmission. By increasing the bandwidth the amount of information to be transmitted per interval increases as well. Bandwidth is measured in Hertz (Hz) or kHz, MHz, GHz. The rate of data transmission is given in Kbit/s or MBit/s. If a transmission in both directions is possible, this is called a duplex connection, otherwise it is a simplex connection. A connection is called half-duplex, if the connection direction can be swapped.

3.5.3 Range

Radio waves are electromagnetic, and they propagate in vacuum with the speed of light. The received power decreases with the square of the distance, which means that the range of transmitters has a principle limit. The realistic distance a radio signal can travel also depends on signal attenuation, also called damping and of course on possible interference. And the range depends on the frequency in use. Signals having low frequency can have a far range even at low intensity. They can also surmount physical barriers, such as walls. This is not the case for signals operating at high frequencies between transmitter and receiver.

Contrary to cable connections, the transmission medium in a radio network does not have visible bounds, which otherwise could be located easily. At the same time the transmission medium is basically unprotected against unwanted signals coming its way. There is also no guarantee that all stations belonging to a wireless network can hear each other at any moment. There is always the possibility that stations may be hidden momentarily. The propagation of radio waves varies over time and is not necessarily isotropic in space. Figure 3.4 from the 802.11 specification of 1999 visualises this effect.

Radio communication, thus, is somewhat less secure and less reliable than cabled connections.

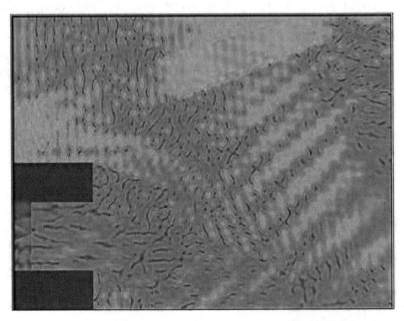

Fig. 3.4. Representative Distribution of the Intensity of Radio Signals (ANSI/IEEE 802.11 1999 Edition).

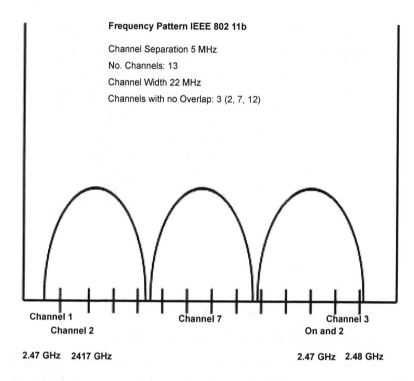

Fig. 3.5. Channel Separation.

3.5.4 Channels

As was already pointed out above, WLANs use the frequency range offered by the ISM band [33]. The 2.4 GHz band is divided between 2.4 to 2.4835 GHz into single channels with a width of about 22 MHz each and a gap of 5 MHz between them (Fig. 3.5 and Table 3.2). Because of spread effects there maybe frequency deviations of 12.5 MHz in both directions against the nominal frequency assigned to the proper channel. Therefore, interference between adjoining channels is possible.

3.5.5 Channel Separation

Interference can be avoided by using only parallel channels with sufficient separation. The best method is to use every fifth channel, which means that only three channels at maximum can be used in a single WLAN. By reducing interference signals transmission power can be augmented.

Table 3.2. Frequencies of Different Channels in the 2.4–2.5 Frequency Band.

Channel	Central Frequency [MHz]	Frequency Spread [MHz]
1	2412	2399.5–2424.5
2	2417	2404.5–2429.5
3	2422	2409.5–2434.5
4	2427	2414.5–2439.5
5	2432	2419.5–2444.5
6	2437	2424.5–2449.5
7	2442	2429.5–2454.5
8	2447	2434.5–2459.5
9	2452	2439.5–2464.5
10	2457	2444.5–2469.5
11	2462	2449.5–2474.5
12	2467	2454.5–2479.5
13	2472	2459.5–2484.5

3.6 Architecture and Components

3.6.1 Data Transmission and Synchronization

In its simplest form data transmission in a network functions via a point to point connection. Two computers are connected via a carrier medium: cable or radio frequency. Each transmission proceeds in three phases: connection set-up, connection control, and connection tear-down. Modems have to synchronize during connection set-up before data can be transmitted. During transmission security mechanisms are active to prevent erroneous transmissions.

Data transmission between sender and receiver is achieved via so called protocols. They control the data exchange in several layers of communication.

3.6.1.1 Networks and Routers

In case more than two participants want to join in, the simple point-to-point connection has to be replaced by a network. In order to locate the different participants they have to have addresses.

In large networks several routes between any two stations are possible. Routers can select the optimal path.

If participants want to use the same carrier medium, both for sending and receiving, special rules to detect or prevent collisions have to be adhered to. The CSMA (Carrier Sense Multiple Access) pools the most popular ones of these.

3.6.1.2 Packet Switching

When telephoning, any two participants are provided with a fixed connection during the duration of the call. This is called line switching. The Internet by contrast works by packet switching. The transmitted information is cut up into blocks called packets.

Every message contains a header with all information required for the data exchange—like the sender and destination address. The packets move autonomously through the net. And, it is quite possible that packets —although belonging to the same message and originating from the same sender—will be conducted to their destination via different routes. Only at arrival will they be patched together again (Fig. 3.6).

The advantage of packet exchange is the efficient usage of existing connections, since the data packets are small and they do not have to queue for long. The communication network is available to all participants. All stations can send packets in turn. Errors will be detected immediately. An erroneous packet will be resent. If a station fails the total message will not be lost completely. The packets choose a different route to reach the destination address.

Data Transmission in Packages

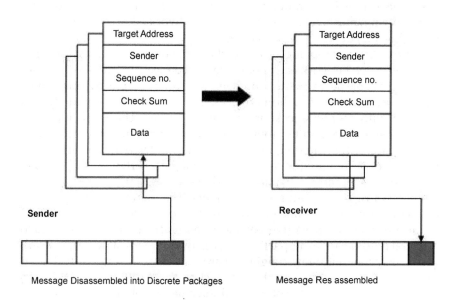

| Message Disassembled into Discrete Packages | Message Res assembled |

Fig. 3.6. Packet Switching.

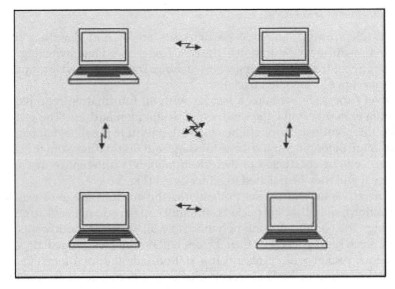

Fig. 3.7. Mesh Network.

3.6.2 Network Topologies

Networks can be built in quite different ways. The main distinctions are between ring, mesh, star, and bus or tree networks. Radio networks use mesh (Fig. 3.7) or star type topologies.

Those different topologies have certain advantages and disadvantages. In a mesh type network each node may be connected with several other nodes. In a cable network this would lead to rather complex wiring, which can be dispensed of in radio networks. The advantage is higher system stability with no bottlenecks in individual cables. The mesh topology therefore stands for a rather robust technology. It has dominated the total architecture of the Internet, even though the latter is not a pure mesh type network but a mixture of various topologies. Partial mesh networks (Fig. 3.8) are alternatives. In partial mesh networks, stations are connected to neighbouring ones, but not to all others. The minimum configuration of a wireless network is one radio cell with two communicating stations.

In a star type network (Fig. 3.9) the individual stations are connected to a central module, for example like a server, by the shortest possible means. In this way the routing is straightforward. The disadvantage is the higher risk of disruption. If the central unit brakes down, the whole network does so at the same time. On the other hand, such networks can be managed by system administrators quite easily, since its configuration can be controlled centrally. In a WLAN, access points take over the role of the central unit, around which the other stations are grouped.

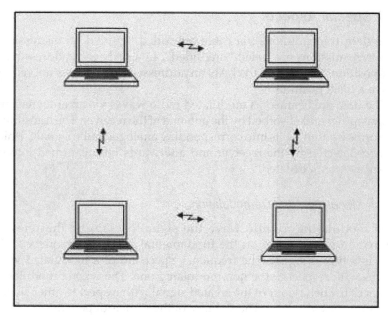

Fig. 3.8. Partial Mesh Network.

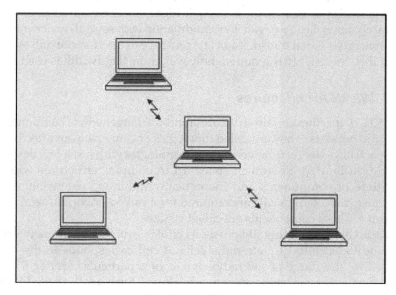

Fig. 3.9. Star Network.

3.6.3 Special Aspects

While data transmissions via cable are called "guided" transmissions, radio transmissions are called "unguided". In a LAN each address has a fixed position, whereas in a WLAN an addressed station does not require to have a fixed position.

The data are beamed as modulated radio waves via an antenna from the transmitter and absorbed by the antenna of the receiver. The transmitter transforms its digital bits into corresponding analogue radio signals, which are then detected by the receiver and afterwards retransformed into bit sequences—demodulated.

3.6.3.1 Modulation and Demodulation

When modulating a radio wave the signal containing the relevant information is impressed on the fundamental or carrier frequency. The result is a mixed signal. The frequency spectrum of a modulated wave changes with respect to the non-modulated one. The type of modulation influences the behaviour of the created signal with respect to other signals in the same environment. The stability of the signals depends strongly on the modulation method employed.

A significant step towards better signal stability was the introduction of the spread spectrum method. Its basic idea is to distribute a signal over several channels, i.e., to spread it via modulation with several carrier waves. This makes the signal transmission less sensitive to interference pulses. On top of this, less energy is required, however, more bandwidth is used up.

3.6.4 WLAN Architectures

IEEE 802.11 specifies certain structures for the arrangement of components in a radio network. They are called topologies or connection architectures. The spectrum extends from very simple topologies with just a few devices covering a limited amount of space up to complex structures with a multitude of components and theoretically unlimited extension. At a minimum, two components are required for a radio network to send and to receive. These components are called stations.

Radio networks generally posses a cellular structure, larger networks are subdivided into discrete radio cells. A cell corresponds to the area covered by the range of the radio signals of a particular sender. These ranges again depend on the antennas employed. Such a cell is called Basic Service Set (BSS) [34].

3.6.4.1 Cells and Stations

A BSS can be illustrated as an oval surface, within which all existing stations can reach each other mutually to exchange data. Precondition is that the stations are within reach of each other and operate at the same channel.

If a station is removed far enough such that no other station can reach it any more, it is outside the radio cell and therefore outside the BSS (Fig. 3.10).

Radio cells may also partially overlap (Fig. 3.11). In this case certain stations are within the range of all others, but some can only reach part of the stations. They remain invisible to the others. Radio cells can be enlarged simply by adding more stations to them.

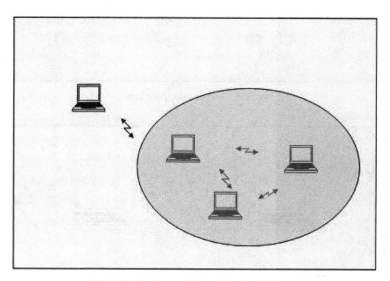

Fig. 3.10. Stations within and without Range.

3.6.4.2 Ad hoc Networks

In its simplest form a WLAN consists of two computers both equipped with a radio component each to send and receive data. If another notebook gets within reach, i.e., within the radio cell, it can participate in the wireless communication. There is no necessity for hubs as in cable networks (Fig. 3.12).

IBSS

As long as such a radio cell consisting of a few computers can operate on its own this constellation is called Independent Service Set (IBSS). Since there are no elaborate preparations to create such a configuration an IBSS

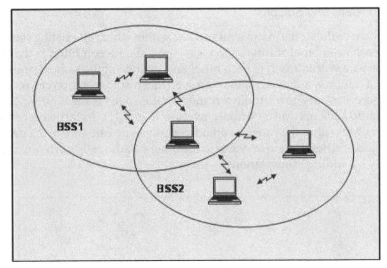

Fig. 3.11. Two Overlapping BSS.

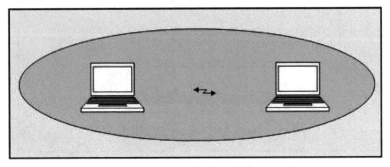

Fig. 3.12. Ad hoc Connection.

[35] is also called an ad hoc network. Within the ad hoc mode all stations are equal. There are no preferred structures or a centre. Data packets are exchanged directly between the individual stations. This type of cooperation is sometimes also called peer to peer workgroup.

In such a WLAN configuration all stations have to have switched on the ad hoc operating mode, and a common transmission channel has to be set up. The range of such ad hoc networks is limited in buildings to between 30 and 50 m. By employing different channels several user groups can constitute separate ad hoc networks not interfering with each other. However, these separate networks cannot get in contact with each other.

Temporary and Spontaneous Connections

The above mode offers itself primarily for spontaneous and temporary networks to organise communication between persons and groups at conferences or fairs, etc. Setting up a network like this is inexpensive since no additional devices are required with the exception of internal and external network adapters. The topology corresponds to a mesh network or a partial mesh network.

Concerning security, this type of network, however, is the least secure. In most cases this operating mode does not permit to activate or configure additional security mechanisms. Attackers just have to adjust to the channel utilised to gain access. Only devices corresponding to 802.11i provide for better protection.

3.6.4.3 Infrastructure Networks

The other operational mode for wireless networks is called infrastructure mode. In this case the Basic Service Set is normally integrated into a larger network structure. The individual stations do not communicate directly with each another, but their information is routed via a router called access point (Fig. 3.13).

The access point functions as a central radio bridge enabling communication between the individual stations. An access point is able to serve a radio cell within a radius between 30 to 150 m. Quite often it not only controls data traffic with client computers acting like a server but also serves as an interface to a cable network.

LAN Portal

Solutions for small businesses or home applications apart—WLANs are deployed mainly as extensions to cable bound LANs. Smooth integration therefore is the general aim and not so much the replacement of LANs by WLANs. However, the emphasis is beginning to shift between the two types of networks with the advent of more powerful and cheaper WLAN components into the direction of WLANs. In many cases an access point serves as a gateway to the local cable network in a company to provide mobile access to data bases or other resources like printers or scanners. In companies or institutions WLAN solutions are commonly used in combination with classical LAN structures or as extensions. The deployment of WLAN components is especially useful in areas, where high mobility of workplaces is required or where cabling is difficult.

In infrastructure mode one single station within a radio cell takes on a dominant role. In its simplest form a WLAN is a radio cell with one access point and several stations. But there are many more variants possible to realize large scale networks.

Distribution Systems

With the help of access points several station clusters, i.e., several radio cells can be interconnected to a so called Distribution System (DS). The allocation of the various stations to the BSS however is—contrary to nodes in a wired network—in principle dynamic. Over time stations can move into the range of a specific BSS and then again out of it to a different radio cell.

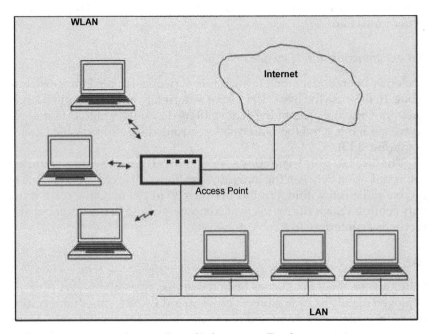

Fig. 3.13. Typical Infrastructure Topology.

Two access points can serve as a bridge between two hard wired LANs. Using powerful directional antennas access points can be used to connect buildings somewhat apart but belonging to the same company.

The next stage of expansion is the Extended Service Set (ESS) [36], where several access points communicate to cover a large area of buildings (Fig. 3.14). Once everything has been configured correctly employees with their notebooks can move freely within a building or between buildings. They are passed on from one access point to the next.

Contrary to the spontaneous linkage in ad hoc mode the infrastructure mode makes the creation of a security architecture possible to prevent unauthorized access to data traffic in a WLAN. The adjustments necessary are made during configuration of the access point. These settings determine the rules, which have to be obeyed by stations wanting to participate in a specific radio cell.

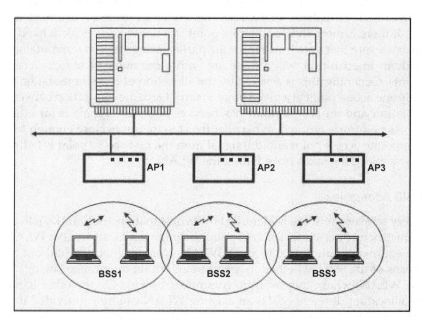

Fig. 3.14. DS with Several Access Points.

Access Points

One single Basic Service Set generally is limited to a circumference of between 30 and 50 m. An intelligent alignment of access points can increase the range of a radio cell to cover distances of up to 100 m. In this setup the access point acts as the centre of the BSS being surrounded by other stations. All these stations have to be able to reach the access point. However, it is not necessary for all stations to reach each other as is the case for an ad hoc network, because all data traffic is always handled by the access point.

Bandwidth Distribution

Theoretically one single access point can administrate a maximum of 2007 stations. But since the stations belonging to one and the same radio cell have to share the transmission medium, i.e., the bandwidth, the effective maximum number of stations is much smaller. In reality this depends on the data volume to be transferred. Practically quite often 20 stations are a useful limit for a single radio cell, if one wants to work with a transmission rate of 11MBit/s. If more stations are necessary additional access points have to be installed.

A WLAN operating in infrastructure mode is by definition more complex than an ad hoc mode solution. Therefore, almost all the security provisions to be realized for a wireless network refer to infrastructure mode.

If there is more than one access point in a network the task at hand is to make sure that data transfer for the participating stations is maintained without interruption, when these are within the range of several access points. Generally, this is assured by the allocation of one single station to only one access point at a given point in time. Once the connection between a station and an access point has been established signals from other access points are ignored. Whenever the station moves close enough to a competing access point and the signal from the first access point is fading a new allocation takes place within the WLAN.

SSID Addressing

Every wireless network is identified by its designation, the SSID [37]. In an infrastructure network all access points belonging logically to that WLAN are endowed with the same SSID. WLAN stations, thus, can find out by means of the SSID, whether an access point within their range belongs to the WLAN, whom they want to communicate with. On the other hand, by allocating different SSIDs an existing WLAN can by subdivided into several distinct networks. This can be useful to separate user groups from one another. All that is necessary is to allocate an appropriate SSID to the stations in question. Stations A, B, and C for example communicate via the access point with SSID "WLANone" and Stations D, E, and F via access point "WLANtwo".

Each of the stations look for the WLAN exactly corresponding with the SSID within the tuning range. Once the correct WLAN is found, the connection is established.

If a station wants to switch between networks, a joker instead of a distinct network designation can be used—normally the designation "Any".

The usage of an SSID has little to do with security since most access points make the chosen name public by broadcasting it. In most cases the administrator can suppress the SSID broadcast by configuring the access point accordingly. This prevents the automatic detection of WLANs around.

When suppressing the SSID broadcast the stations have to know the designation beforehand, if they want to hook up to an access point. But this again is no major obstacle for adequate sniffer tools as long as data are transmitted without encryption.

3.6.4.4 Mobil Internet Access

Current developments have made smaller WLAN solutions also attractive for private applications in households. This is partly owned to the fact that fast DSL connections can be utilized from different locations without having to change any cabling. The necessary WLAN components have to combine with a router or with DSL modems having router functionality integrated (Fig. 3.15).

This solution is preferable to the Internet Connection Sharing (ICS) [38] because a particular computer does not have to be active to set up an Internet connection.

Access Point as Router

To provide the stations of a radio cell with a common Internet access the access point has to be able to act as a wireless gateway. In this way the access point takes over router and DHCP server functions. For this normally the Network Address Translation Protocol (NAT) is used to hide the WLAN behind a single IP address against the outside world.

IP Address Allocation

With the help of routing functions data can be transferred from a radio cell to the IP addresses in question to the Internet or vice versa. DHCP takes care of the automatic allocation of the IP addresses for the individual stations

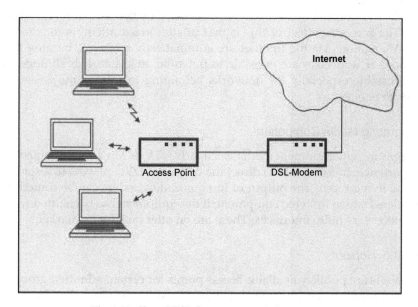

Fig. 3.15. Shared DSL Connection with the Internet.

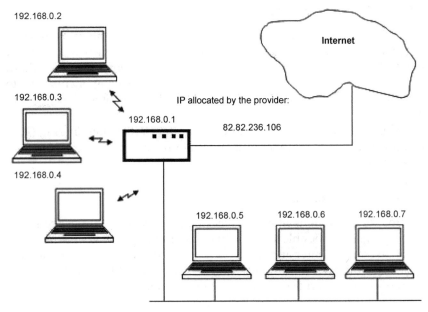

Fig. 3.16. Usage of a Common IP Address with NAT.

within the same radio cell. By employing NAT the IP address for Internet access allocated by the provider is mapped to the individual IP addresses in such a way that several stations can use the Internet access directly even if only one single IP address has been allocated by the provider (Fig. 3.16).

The nice side effect of this is that unauthorized attempts to contact WLAN stations via the Internet are automatically inhibited, because the stations IP addresses are invisible to potential attackers. This solution is comfortable especially for networks belonging to small enterprises or private persons.

Integrating Media Components

Integrating entertainment media, such as PC, TV or Music have gained in importance. It is possible to direct the output of DVD players to a screen in the living room. The output of Internet radio senders can be routed in wireless fashion to stereo equipment, if the equipment has been fitted with the necessary radio interfaces. These are on offer now in the market.

3.6.4.5 Hotspots

Hotspots are publicly available access points for certain admitted groups. Typical locations for hotspots are airports, railway stations, hotels, fairs, congress centers, cafes, and public libraries providing wireless Internet access or other services to be utilized via notebooks or smartphones.

Hotspots have been widely accepted by now. The access procedure has to be as simple as possible—charges apart. Generally a GUI is presented to the user, where he can login and the charging process is triggered off, unless the service is free of charge. In most cases access codes are necessary to clear the way into the Internet. These codes are handed out via prepaid cards or voucher.

Hotels offer prepaid cards valid for 24 hours. The code can only be used once. Codes are also distributed via SMS. Charges are billed to the mobile phone account.

Hot zones are areas where several hotspots cover them by floating intersections. The open source movement is to promote these freely available networks. Such self-organizing networks are on offer in large capital cities, but they are also available more recently between villages in India.

3.6.4.6 Open Networks

These community networks are special cases in terms of security, because they are conceived to make available data freely and openly. As long as these networks offer information via established hotspots there is no need for security procedures, such as encryption or address filtering. To make access as easy as possible a unique designation for an SSID is dispensed of and the joker name "any" is used.

3.6.4.7 Roaming

Roaming, is when a mobile phone moves out of its network region and attaches to different networks in order to resume service especially when travelling abroad. The guest access to a different network happens without notice, but becomes visible through higher charges on the phone bill later.

The term roaming is applied, when mobile users can change without problem from the receiving range of one access point to the receiving range of a different access point in large infrastructure networks. The handover should be completely transparent if possible, especially without interrupting a radio connection to avoid potential data loss during a running transmission.

Roaming is possible through the regular broadcast of beacon frames emitted by an access point to indicate its existence within a radio cell. WLAN stations scan all available channels to locate beacon frames in regular intervals unless they have already been allocated to a specific access point. If a station receives several beacon frames at the same time the signal power decides, which access point has to be selected. But as soon as a station separates from access point A and approaches access point B the connection changes without notice by the user and without impact on the currently running data transfer.

Several access points thus can cover a larger area with overlapping radio cells permitting network access anywhere. Users in a big hotel move around from one conference room to the next without losing their connection to the network. As long as activities take place on the physical layer with separate channels this will work—not excluding the occasional odd problem.

Since the number of possible channels is always limited, a special Spatial Division Multiple Access (SDMA) [39] procedure is available to protect against possible interference. This procedure permits the re-utilization of channels, once radio cells are at a certain distance from each other.

Roaming is of special importance in connection with public hotspots. A single login would provide for movement between different hotspots. Precondition is the harmonization of access procedures.

3.6.5 WLAN Components

In this section we discuss those components, which are required separately from a device, which is not already integrated. One has to differentiate between those components, which are required for a client (adapters) and those required for setting up of the proper network (access points, routers). Adapters can be classified according to the respective clients:

- Laptops
- Desktops.

Further differentiation depends on the interfaces.

Access points are either singular components or are integrated into a router solution, if more than one network interface is required.

When buying WLAN components certain criteria have to be observed with respect to security and performance, as follows:

- Number of channels configurable
- Configuration of SSID, deactivation of broadcast possibility
- Encryption procedure provided
- Compatibility with the IEEE 802.1x standard
- Possibilities of address filtering
- Possibility to use an Access Control List (ACL) [40]
- Compatibility of authentication method between access points and clients
- Additional mechanisms for access control.

3.6.5.1 Adapters

Adapters for Mobile Terminals

Here we will only consider laptops/notebooks. In the most favourable case the device is already equipped with a WLAN adapter. No extensions are necessary. Otherwise adapters are of the following type:

- WLAN USB adapter
- WLAN cardbus adapter.

WLAN USB Adapter

This adapter requires only a single piece of cable. It connects with a laptop via a USB plug. The performance of the adapter depends on the throughput limits of the USB interface.

WLAN Cardbus Adapter

A cardbus adapter allows for higher processing rates (Fig. 3.17). The flat card is inserted into the appropriate slot at the laptop. Only the last inch of the card remains visible displaying two flashlights. Insertion should take place only after the complete boot of the computer.

Fig. 3.17. Wireless Notebook Adapter from LINKSYS.

Adapters for Desktops

Adapters for stationary terminals (PCs, desktops), so called PCI adapters (Fig. 3.18), must be fixed into the device. Preconditions are:

- Free slot available
- CD-ROM drive

Fig. 3.18. Wireless PCI-Adapter from LINKSYS.

- 500 MHz processor
- 128 MB RAM
- Windows 98 upwards.

3.6.5.2 Access Points

WLAN Access Point

This device fulfils all requirements to set up a simple WLAN with clients configured appropriately. Apart from this an access point does not offer any other functions, such as Internet access.

WLAN Router

Just like classical routers, a WLAN router (Fig. 3.19) manages the control of multiple incoming and outgoing signals using parallel channels. So Internet and WLAN interfaces can be handled at the same time—even an additional telephone switch, Fax, printers, a classical LAN and of course a PC or a laptop as clients. A router is equipped with an antenna, which can be adjusted manually.

Fig. 3.19. Wireless Router from D-Link.

3.6.6 Configuring a WLAN

Before configuring a WLAN certain criteria have to be taken into account at the planning stage, such as:

- Decision about the authentication method
- Shielding against other technical devices emitting electromagnetic waves
- Preparing a layout that avoids radio dead spots
- Avoidance of channel overlap by multiple users.

The configuring of a WLAN and its components proceeds in three distinct phases:

- Drivers
- Hardware
- Proper configuration.

Configuration procedures differ depending on whether the object is just a simple adapter or a complete infrastructure network. Within the network routers and clients have to be setup as well. In all cases detailed instructions for installation—normally on a CD—are usually provided by the manufacturer guiding the user through his options. At this point we will not repeat the user guidance in detail.

3.6.6.1 Drivers

For all adapters driver installation is necessary. The drivers are provided with the adapter on a CD or can be downloaded from the Internet. After loading the usual graphic user prompting takes place. This and all other installations should proceed under the administrator account. In this way the driver is saved securely on a path decided on by the administrator. This is the only substantial decision required. After finishing the installation the computer has to be rebooted.

3.6.6.2 Hardware and Configuration

WLAN USB Adapter

This is how to proceed:

- Insertion of setup CD and start.
- Connecting USB plug of the adapter to a USB port available.

After this the user is asked to select an "available wireless network" by means of a choice menu presenting all SSIDs currently available. The user selects the network, with which he wants to work and establishes the connection via "connect". If the target network is not in the list, it can be added.

If the network is to be protected by WEP, the WEP key belonging to the network is requested. Only after entering this key a connection to the network is possible. If protection is controlled by WAP a special passphrase for the net is required as input.

WLAN Cardbus Adapter

This is how to proceed:

- Insertion of setup CD and start
- Introduce the adapter into the cardbus slot
- Start installation.

After this the user is asked to select an "available wireless network" by means of a choice menu presenting all SSIDs currently available. The user selects the network, with which he wants to work and establishes the connection via "connect". If the target network is not in the list, it can be added.

If the network is to be protected by WEP, the WEP key belonging to the network is requested. Only after entering this key a connection to the network is possible. If protection is controlled by WAP a special passphrase for the net is required as input.

PCI Adapter

This is how to proceed:

• Insertion of setup CD and start.

After positive confirmation and execution of the following steps the computer shuts down. Now the manual assembly of the adapter can start.

The PC encasing has to be opened to find a free slot on the motherboard. The adapter is placed on the PCI slot and its ear fixed with a screw to the PC frame. After this the housing is closed and the antenna attached to the adapter. Now the computer can reboot.

After this the user is asked to select an "available wireless network" by means of a choice menu presenting all SSIDs currently available. The user selects the network, with which he wants to work and establishes the connection via "connect". If the target network is not in the list, it can be added.

If the network is to be protected by WEP, the WEP key belonging to the network is requested. Only after entering this key a connection to the network is possible. If protection is controlled by WAP a special passphrase for the net is required as input.

Access Point

Initially all default settings by the manufacturer have to be replaced by individual security settings to prevent unauthorized access.

Connection of an access point proceeds either directly to a PC by a crossover cable or via a switch or hub in a hardwired network. Switch on access point and start installation CD. The setup assistant searches the net for the access point. Depending on the manufactured type either a new SSID or an automatic assignment of an IP address are possible.

Router

Precondition: modem or DSL with corresponding telephone connection.

Preparation:

• Connect power supply
• Connect one of the five output sockets of the router via a patch cable to the computer
• Connect WLAN interface of the router to the modem.

This is the usual process under Windows ©:

Select "System Control" via the start menu of the computer, select "network connections", then "LAN connection". In the following menu select "Internet protocol (TCP/IP) and properties". Only one single IP address can be obtained automatically or entered by choice.

From now on (for some providers right from the outset) user prompting continues via the Internet browser in the manufacturer's domain. Normally the included IP address has to be entered. The following parameters have to be adjusted:

- Router password
- Internet connection type (DHCP, static IP, PPPoE)
- Network mode
- SSID designation
- Radio channel
- WEP or WAP securities.

3.7 Security Requirements

The peculiarities of radio transmissions require special measures to secure wireless communication in local networks. Before detailed procedures from relevant standards are presented, a brief summary about the objectives is given as follows:

3.7.1 Assuring Availability

The sensitivity of wireless connections against interference is a general problem. If other technical devices transmit within the same frequency spectrum as WLAN components, WLAN communication will be disturbed or even prevented. These other devices can be microwave ovens, surveillance cameras, or Bluetooth traffic. It is also quite possible that an attacker tries to deliberately interfere with WLAN transmissions.

One of the reasons for careful planning and implementing a communications network is to achieve maximum availability. One aspect is the optimal placement of components fixed in locations, i.e., access points. Another important variable is the selection of the most favourable operational modus and with this the decision about possible frequency ranges and transmission rates. Because of the sensitive nature of the network constant observation of its performance is necessary to identify the causes for any malfunctioning.

3.7.2 Assuring Data Integrity

In any network—be it wireless or cabled—data have to reach there destination complete and unaltered. Once data have been manipulated on their route, the receiver has to be enabled to detect this fact to be able to react. The effect is the same, whether the manipulation has been done on purpose or whether it was only the result of a transmission error.

3.7.3 Assuring Authenticity

Both sender and receiver of messages have an interest that the authenticity of the other side is guaranteed. Special access controls have to make sure that sender X cannot pose for sender Y. The same goes for the receiver. This is most important for legally binding transactions including commercial orders, invoices, etc.

3.7.4 Assuring Confidentiality

In contrast to communication in public networks, in which publicly available information is offered to anyone, who wants it, confidentiality plays an important role in private wireless networks with respect to data protection. The desired levels of confidentiality have to be implemented in practice. Since radio signals can be listened into the only solution is encryption. Encryption has two objectives:

- To protect transmitted information and
- To protect link data.

3.7.5 Security Risks

The fact that—so to speak—wireless transmissions use free space, this transmission medium makes it easier to eavesdrop to them than to data transmitted via cabled connections. The security requirements therefore are different with respect to LANs [41]. On top of that LANs are locally fixed with known users. WLANs neither have visible geographic boundaries nor is it apparent, who is currently connected.

These are the most important motives for attackers and their most common forms of attack:

- Technical challenge: playful hackers just wanting to find out, whether they can get access somewhere without any intention to wreak havoc;

this may include the intention to listen in without the knowledge of other people and to penetrate their private spheres. Tools can be obtained via the Internet.

- Criminal intention: the purpose is to damage other persons or organizations or to enrich oneself.
- Unauthorized sharing of Internet access: if a WLAN is operated in combination with DSL access, someone could try to use the Internet access without authorization. There exists the possibility to misuse some ones account for downloads of confidential data or for criminal contacts.
- To gain direct material advantage: all kinds of unauthorized accesses are imaginable without knowledge of the party concerned even over a prolonged period of time.
- Insertion of data and software via an unauthorized station in a WLAN to drop off selected data. This is done by pretending authorized identity to an access point. Examples: implantation of spyware [42], spying data of credit cards, attacks by Trojan Horses [43] to steal important company data; viruses to destroy data.

3.7.5.1 Spying

This is the prime motive for penetrating unprotected WLANs. WLANs are easy to detect, since they broadcast beacon frames into the surrounding space to attract attention. An attacker on the street in his car can collect these frames via an antenna and will gratefully enter this WLAN community as a blind passenger. It is sufficient that he approaches the building in question close enough with his notebook or smartphone without being noticed in the first place. With tools obtainable anywhere, the WLAN check is done and the names of the access points identified. Sniffers [44] can thus enter the range of a radio cell without being detected, collect all data traffic and analyse it.

With special tools, the number of WLAN users can be determined and checked whether the network has any protection at all. Estimates say that the majority of WLANs in operation today is not sufficiently protected. In many cases an intruder can enter a radio cell unchecked.

3.7.5.2 Decoding

Even if a WLAN employs encryption techniques this may not suffice to prevent illegal recording of data. Once the attacker has registered enough data packets he can try to break the cipher key with statistical methods. For example, to break a WEP key, data recording of only a few hours usually suffices. There are tools available on the Internet, which have been developed for the first generation of WLAN components, where encryption mechanisms have been weak. Companies can use such tools themselves to discover weak points in their own WLAN configuration.

WLAN protection at a certain level quite often is not adequate against determined attackers. Even after initial failure more powerful tools will be fielded to arrive at the intended end. Early discovery and continuous observation are therefore necessary.

An additional danger is posed, if an attacker tries to pretend a legitimate user identity with the help of network addresses used in the WLAN and thus obtain access to protected data areas.

The fact that WLANs are prone to spying in one or another way calls for appropriate countermeasures. Part of the strategy can be a WLAN trap to find out, whether attacks from outside are attempted. The euphemism for this is "honey pot network" [45]. Companies can configure a less protected part in their otherwise well shielded network using dummy data for fictional business traffic. By analysing log files of the access point illegal attempts can be filtered out.

3.7.6 The Physical Layer

The problem to be solved on this layer concerns the fact that within the frequency range provided many stations may want to transfer data at the same time. This happens, when the range of potential senders and receivers overlap.

3.7.6.1 Spread Spectrum

To be able to distinguish between the different participants FHSS and DSSS are employed to generate the necessary signals. FHSS is the older method, whereas DSSS is used more widely today. Both procedures are not compatible. This means that within a WLAN all components have to follow either one or the other.

The spread spectrum procedures are less sensitive to interference or electronic disturbances than single channel methods.

FHSS

The Frequency Hopping Spread Spectrum (Fig. 3.20) was originally developed during the Second World War to control torpedoes. As indicated by its name the radio signal is subdivided into small segments and hops within fractions of a second several times from one frequency to another. The quasi random selection of the frequencies is achieved by the Gaussian Frequency Shift Keying (GFSK) [46] method. To be able to handle the signals the receiver has to know the pattern of these hops.

Other sender/receiver pairs may at the same time use a different hopping, so that several data transmissions can run at the same time without interfering with each other.

If in the unlikely case a collision takes place, the system sends the data packet again, until the receiver sends back a hand shake. The 2.4 GHz band for WLANs is subdivided into 75 subchannels with a width of 1 MHz each. The disadvantage of FHSS is the relatively high overhead with respect to the useful data generated by the frequency hops. Therefore, data transmission with FHSS is relatively slow. The maximum is 1 MBit/s.

DSSS

The Direct Sequence Spread Spectrum procedure is widely used and works differently from FHSS. DSSS uses just one single 22 MHz wide channel without frequency hopping. The data stream is combined via an XOR operation with a so called chip or chipping code. The data is represented by a random sequence of bits known only to sender and receiver (Fig. 3.21). The zero is represented by inverted chipping code.

The code spreads the transmitted data over the available bandwidth. Longer chips need a wider bandwidth, but increase the probability of the data being transmitted correctly.

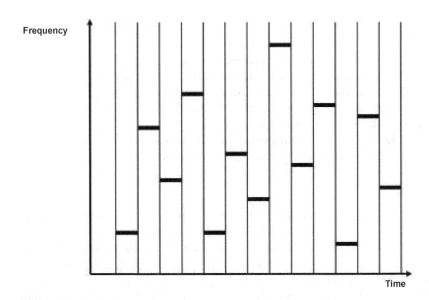

Fig. 3.20. Frequency Hops.

The advantage of DSSS is that the receiver can easily find out, whether the data have originated from the same sender which has generated the code. The procedure also facilitates error checking, since bit patterns not corresponding to the code can be filtered out. If there are one or two bits in the pattern, which have not been transmitted correctly, an automated correction takes place without having to send the data again. Since the protocol overhead is less than that for FHSS higher transfer rates are possible. Newer variants of 802.11 take that into account.

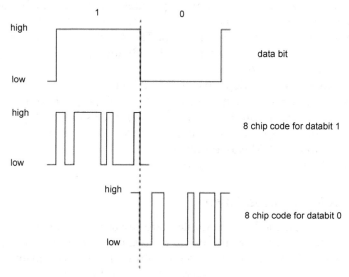

Fig. 3.21. The DSSS Procedure.

With DSSS the data packets are extended by a 144 bit prefix. 128 bits are used for synchronization and 16 bits for a start of frame field. This is followed by a 48-bit header with information about the transmission rate, the length of information within a packet and a control code. Since only the header fixes the transmission rate of the succeeding user data, the header itself is always transmitted with 1 MBit/s beforehand.

Even though the prefix will be removed in the further process its length is still taken into account, when calculating the transmission rate. The effective transmission rate, thus, is always less than the nominal rate.

HR/DSSS

An improved option of DSSS is called High Rate Direct Sequence Spread Spectrum (HR/DSSS). This is the frequency spreading procedure most commonly used in WLANs. Just as DSSS itself HR/DSSS operates within the 2.4 GHz band. It uses a modulation technique called Complementary Code Keying (CCK) [47]. Data can be transmitted at a rate of up to 5.5 or 11 MBit/s.

OFDM

Another option for signal generation in radio networks is called Orthogonal Frequency Division Multiplexing (OFDM). This technique operates within the 5 GHz band. Contrary to FHSS and DSSS, OFDM transforms the digital data into multiple analogue signals in parallel. The frequency bands are separated into four channels each, which are split again into another 52 subchannels of 300 KHz width each. The subchannels may overlap. Interference is avoided by scheduling. The advantage of this procedure is a much higher transmission rate.

The transmission rate depends on the modulation technique employed. With Binary Phase Shift Keying Modulation (BPSK) [48] 6–9 MBit/s can be obtained, with Quadrature Phase Shift Keying (QPSK) 12–18 MBit/s and with Quadrature Amplitude Modulation (QAM) 24–36 MBit/s.

Furthermore, OFDM was applied to the 2.4 GHz frequency band, with the aim to obtain similar transmission rates here as well.

3.7.7 Medium Access Layer

The procedures specified by WLAN standards for accessing the transmission medium generally differ from those in the 802 family defining accesses in LANs. These differences are due to the nature of wireless transmissions. The 802.11 standard defines different services for the carrier access layer controlling data exchange within the network. These services concern the way, in which data are prepared for transmission and the necessary security provisions.

3.7.7.1 Frames and Fragmentation

If data are to be exchanged via a wireless connection, they have to be divided up and packaged in a suitable way. The 802.11 standard calls these MAC protocol data units (MPDUs) [49]. It defines exactly how such a data packet or frame should look like. There are different types of frames: data frames for user data, control frames for control data and management frames to manage the network operations.

A data frame consists of a MAC header, the frame body containing the user data and a frame check sequence, containing a check sum as a 32-bit cyclic redundancy code (CRC) [50]. Figure 3.22 shows the components of a data frame.

Fig 3.22. Components of a Data Frame.

All other fields and their sequence are fixed, however, the length of the frame body can vary. The first address contains the target address of the final destination, the second address is the original address of the sending station, and the remaining addresses control the forwarding.

When configuring an access point a fragmentation threshold can be defined. This value determines the maximum packet length. If this length is exceeded the packet is divided up into so many more fragments.

3.7.7.2 Avoidance of Collisions

The MAC layer controls primarily the data traffic within the radio network. 802.11 defines several methods for that purpose. One aspect is the avoidance of collisions. Collisions would otherwise occur, when several stations are sending data packets at the same time.

CSMA/CA

The access to a radio channel in systems according to the 802.11 standard is controlled by a random procedure called Carrier Sense Multiple Access with Collision Avoidance (CSMA/CA) [51]. This technique permits the simultaneous access of several devices to the carrier medium. The working of CSMA/CA can be described in the following way:

When a station wants to send a data packet it checks the carrier medium, whether other signals are present. If it cannot detect any signal it waits for a short random time interval (inter-frame spacing) and then checks again, if the medium is clean. If this is the case the data packet is transmitted. The station having received the packet checks its integrity. If everything is okay, a receipt is sent after another short inter-frame spacing. If the sending station does not receive an acknowledgement, it is assumed that a collision with a different data packet has taken place. The station waits again for a random time interval and then tries again.

RTS/CTS

An optional extension of CSMA/CA is RTS/CTS [52]. This procedure is applied to master the problem of "hidden" terminal devices. These are devices, which sometimes cannot be reached because of signal attenuation.

Initially the station which wants to send emits a request to send a packet to reserve a transmission channel. The receiver acknowledges this reservation with a clear to send packet. All other stations remember the holding time established by the RTS/CTS packets and refrain from sending data themselves during this time interval. If more than one station tries to send data at the same time, CSMA/CA instructs all other stations to refrain and try again later.

While configuring an access point one can normally define a threshold. This value determines, whether the packet transmission can be handled by the CSMA/CA or by the CSMA/CD [53] method otherwise used in LANs as a function of packet size. In the latter case, the packet will be sent after a certain holding period.

3.7.7.3 MAC Addresses

The protocols of the 802.11 MAC layer operate within the same address space generally provided for local networks adhering to the 802 standards family. To identify components the Medium Access Control (MAC) address is used. This is a unique 48-bit serial number assigned by the manufacturer of the relevant network component. The first 24 bits contain the manufacturers ID, assigned by the IEEE, the rest is filled up by the manufacturer himself. This number is generally represented in hexadecimal code like 00-09-5b-e7-b3-5c with a hyphen as separator.

MAC Addresses and IP Addresses

With the help of MAC addresses the MAC layer can get in touch with higher levels of the ISO model. In this way it is possible to assign for example an IP address to a component with a specific MAC address (Fig. 3.23). This mapping is done by the Address Resolution Protocol (ARP) [54].

#	IP Address	Device Designation	MAC Address
1	192.168.0.3	MYTRAVELMATEXP	00:00:e2:30:6e:82
2	192.168.0.4	DELLPROF	00:10:5a:bb:0b:cb
3	192.168.0.5	FUJI	00:30:f1:15:4b:6f

Fig. 3.23. Device List in an Access Point with the Mapping for IP Addresses to MAC Addresses.

Since both LANs and WLANs use MAC addresses in the same way it is no longer possible to distinguish, whether a user utilizes a LAN or a WLAN component at the Internet protocol level.

Address Filtering

Access points have the option to have special MAC address filters configured, so that only stations with defined MAC addresses will have access, which is denied to others. For this purpose tables have to be manually administrated, which can be tedious in large networks. On the other hand, one of the known security problems is the fact that MAC addresses can be faked. When an attacker finds a registered MAC address, he can program his own device to show just this address to the network. This is called MAC address spoofing [55].

3.7.7.4 SSID Network Name

Each wireless network can be identified by a network designation. This name can be chosen arbitrarily and is called the Service Set Identifier (SSID). Its length can be up to 32 characters. This value can also be set to "0" (zero) corresponding to operation mode "any". In this case any station can hook up to the access point. The access point broadcasts this mode at regular intervals as beacon frames to attract attention. In case the network has a proper designation, the access point waits for the polling of stations, to which the SSID is known.

In this way the SSID can be used to control access. But this mechanism is of limited value, because the SSID designation is transmitted unscrambled and can thus be easily detected. This is particularly precarious once the designation allows for conclusions as to the name of the user or his company. This is to be avoided.

3.7.7.5 Authentication Procedures

It has already been mentioned that the nature of wireless networks and their mode of connection makes them particularly prone to data theft and spying. To impede this several authentication methods have been defined on the MAC layer. Before a station can communicate with a WLAN, authentication has to take place to verify the identity of the station as being a registered member of a group of stations constituting the network. There are two types of authentication: Open System and Shared Key.

Open System [56]

This first type being the default basically dispenses with a proper authentication and is, therefore, called "null authentication". Every station requesting an authentication of this type gets it from every other station configured in the same way. The procedure works in two stages. At first the authentication is requested and thereafter confirmed, if it works out to be correct. Only thereafter communication within the WLAN is granted. As long as all the components within the network operate with open system authentication, any notebook can share messages with all reachable networks as long as they are not encrypted.

Shared Key

The Shared Key authentication can only be applied, when the Wired Equivalent Privacy (WEP) mechanism is activated. This procedure requires that the station and the access point own the same key. The station has to prove to the access point that it knows the key indeed. This happens by sending a test piece. This operation proceeds along the following line:

The sending station emits an authentication request to the access point. It includes its own MAC address to identify itself and an Authentication Algorithm Identification (AAI), which controls the authentication method —in this case a "1" for Shared Key and a sequential number, controlling the sequence of the four authentication steps.

The access point replies with the same AAI, adds first 1 to the sequential number and then a random number of 128 bytes in length.

In its turn the station encrypts all three elements with a shared key, while increasing the sequential number and returning everything to the access point.

The access point checks the reply and decrypts the test message. If it matches this is proof that both cipher keys correspond. The access point sends an acknowledgement to the station and permits access to the network.

This procedure allows a well directed message exchange within a specified sample of participants.

Wired Equivalent Privacy (WEP)

From the beginning the standard IEEE 802.11 provided for security architectures to deal with the sensitivity of wireless connections. WEP offers a basically symmetrical encryption procedure to deny unauthorized access to sensitive data by trespassers. The secret cipher key is simply distributed between access point and adjacent stations. How this should happen is left open by the standard. This means that within the same WLAN only a single shared key is used.

WEP can be used only once to encrypt transmittable data packets. In this case the Open System Authentication is used. The sending device encrypts the data with the configured cipher key. The receiving component uses the same cipher key to decrypt. Another variant combines WEP encryption with Shared Key Authentication.

Today, WEP is heavily criticized, because the procedure does not bear up against serious attacks. There are programs, such as AirSnort [57] under LINUX and later WINDOWS that are capable to uncover keys used by WEP. A sufficiently large portion of data packets (5–10 million) has to be scanned to do the trick.

Stream Ciphers

The reason for the weakness of the WEP mechanism is basically the choice of the encryption procedure. The cipher key lengths in question are 64 and 128 bits with a 24 bit default. The user can thus only work with 40 and 104 bits resp. (This is why manufacturers talk about 40 and 104 bit encryption.)

With 40 bit generally four keys are supported, which are generated by five groups of two hexadecimal values each like: Cd 55 63 EF 56. The cipher keys are either entered manually or created automatically. In the latter case cipher key generation is associated with a password.

WEP uses the RC4 [58] algorithm (Fig. 3.24). RC4 stands for Rivest Cipher No. 4, pointing to developer Ron Rivest, who also participated in the development of the RSA encryption method. This cipher is also called stream cipher—a random generator, which permits to generate a stream of ciphers of any length from a single secret cipher key with fixed length.

The secret cipher key is combined out of a random initialization vector (IV) of 24 bits length and the 40 or 104 bits reserved for the access point. Before sending user data with a message a check sum of 32 bits length of the non-encrypted data will be created. This Integrity Check Value (ICV) is attached to the data. The cipher stream is generated with exactly the length corresponding of the adjusted user data length. Since the maximum length of such messages is limited to 2304 bytes the frame body can attain up to 2312 bytes with WEP.

Cipher stream and enlarged user data will be linked bit by bit to each other by XOR operations. The result is transmitted preceded by the IV. The receiver inverses this procedure and reconstitutes the original contents of the message. After decryption the check sum is generated once again and compared with the original value. Once the result is okay the data packet is accepted otherwise rejected. WEP only encrypts user data and the check sum but not management or control data.

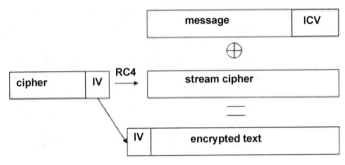

Fig. 3.24. The Functioning of WEP.

Lack of Cipher Key Administration

Another weak point of WEP is cipher key administration or rather the lack of it. WEP only uses one single cipher key for all components in the same WLAN. This quite often means that secret cipher keys are never changed or only very rarely. And guest users need to know the cipher key as well to have access to the WLAN. On top of this some manufacturers of WLAN adapters store the key in such a way that it can be uncovered quite easily. Once a key has been uncovered or corrupted, because an untrustworthy person knows it, the whole network is in danger.

Many WLAN components actually in use still offer WEP in spite of its known weak points. Therefore, one should at least ameliorate the situation by changing the WEP cipher keys more frequently.

Insufficient Cipher Key Length

The basic weak point of WEP was initially its 64(40) cipher key being to short to withstand attacks. Such cipher keys can be identified rather quickly with the assistance of programs by testing all possible bit combinations for the recorded messages under scrutiny. Quite often manufacturers aggravate this problem by generating the hexadecimal key only out of ASCII strings, reducing the number of variations even more. This brute force attack is now much more difficult for 128 (104) bit cipher keys.

Initialization Vector

Another serious weak point is the short length of the 24 bit initialization vector. The initialization vector is generated by the sender and should be different for each transmittable data packet. Components manufacturers should make sure these possibilities exist. This is not always taken care of, which means that the soft spot regarding the initialization vector persists.

Stream cipher operations can only be secure once the bit stream generated by the relevant algorithm differs between two data packets. With 24 bits a maximum of 2^{24} corresponding to 16.8 million cipher keys can be generated. Since some manufacturers only provide simple counters for the initialization vector the spectrum of randomly generated combinations is reduced significantly. But even, if a cipher key has been generated completely randomly the probability that a cipher key has already been used before is > 50% after the transmission of 4823 data packets.

If there are two data packets within the same data recording using the same cipher key, an XOR operation applied to both encrypted texts could eliminate the cipher stream and reconstitute plain text. This is the entry point for decrypting other plain texts as well. This opens up the possibility to introduce data packets until the next cipher key change since the access point regards them as correctly encrypted.

Unreliable Authentication

Even the authentication protocol described above can be broken in a similar fashion, once an attacker records it, since both the authentication procedure and the user data use the same key. Another weakness in this procedure is its one-sidedness. A station has to prove its identity to an access point but not vice-versa. Therefore, a station does not have the possibility to find out, whether the expected access point is an unfriendly camouflaged access point.

3.7.7.6 Better WEP than No Protection at All

Even though WEP security can be broken by suitable means within a span of hours, an improved cipher key change frequency ameliorates the situation somewhat. Attackers would have to start over again to break the encryption. But in practice many users do not even use the insufficient WEP provisions since they are only optional.

The susceptibility of WLANs has been demonstrated by so called war walking. Someone tries to detect unprotected WLANs via notebooks or smartphones from the outside. The intention is to use Internet access free of charge, spying on other people and to manipulate foreign data.

3.7.7.7 WPA and WPA2

To compensate for the weaknesses of WEP proprietary mechanisms like WEPplus or Fast Packet Keying [59] have been developed to implement better security procedures. The Wi-Fi Protected Access (WPA) procedure introduced by the Wi-Fi Alliance in 2002 has gained wider circulation.

TKIP

WPA uses a procedure called Temporal Key Integrity Protocol (TKIP) [60] developed by the Task Group for the later standard 802.11i. For reasons of downward compatibility the encryption algorithm RC4 has been maintained, but instead of a permanent cipher key temporary keys have been introduced. TKIP, thus, is basically an improved variant of WEP, comprising an extended initialization vector, dynamic cipher key generation and a cryptographic Message Integrity Check (MIC) also called "Michael".

3.7.7.8 802.1x [61]

Another option is 802.1x. This standard specifies user authentication und administration of cipher keys for LANs generally. For this purpose a special authentication server is necessary taking over the overall access control of the whole network. On this basis separate clients and the server exchange messages to mutually identify themselves. Before a successful completion of authentication has not taken place, neither the access point is able to accept data from a station, nor can a station receive data from the access point. As a side effect, one off session cipher key is generated to protect the succeeding data traffic.

Authentication Procedure

802.1x offers a convenient way to establish cipher keys for WLANs to provide for typical encryption security. The authentication is access-based in mobile Wi-Fi networks. Authenticators, integrated in wireless access points, are elements of the authentication system, which can grant or deny access. The user tries to get access to the WLAN via the authenticator, which in turn demands certification of credentials for authentication at the access point. Authentication servers belong to the RADIUS (Remote Authentication Dial-In User Service) server family.

The 802.1x process (Fig. 3.25) starts, when the user gets in touch with the authenticator to gain access to the network.

The system blocks all traffic of the client except what is necessary for authentication. Once the access attempt has been detected the authenticator demands the identity of the user.

The user identifies himself and his request to gain access to the authentication server via the authenticator. Thereafter, an exchange of question and answer messages take place with regard to the protocol. Sometimes the network not only authenticates the client, but the client itself tries to find out, whether the network it is trying to gain access to is trustworthy itself. This increases the question and answer traffic.

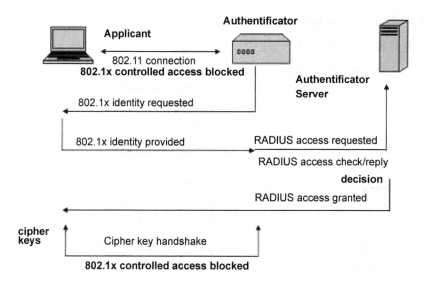

Fig. 3.25. The 802.1x Procedure.

Once the user has met all conditions the server informs the authenticator to grant access. To assure security user and authenticator exchange cipher keys.

Typically, the user authenticates himself via one of the many EAPs (Extensible Authentication Protocols) [62].

IEEE 802.1x generates single user cipher keys for each session. This is to prevent problems related to some security techniques, such as WEP, which need a decoding key for all terminals connected to the same access point.

When the user logs off, he sends a message to the authenticator, which resets the access to be blocked for any non-authenticated traffic.

Up to today the standard 802.1x is not used extensively despite all its advantages. Reasons are to be found in relatively high investment and running costs. Organizations wanting to deploy the standard have to buy additional components and the RADIUS server, install, and configure them. Many users are not used to that. The complex installation process costs time and money. On top of this the suppliers quite often provide their specific version of the standard, which are not always compatible between them. 802.1x can handle so many different authentication methods that users have problems to select the most advantageous.

The Open Sea Alliance [63] tries to solve these problems by developing open source reference installations to reduce the dependency on providers and operating systems. In this way they hope to obtain better interoperability and a wider use of the standard.

3.8 Recent Developments

3.8.1 White Spaces

The current frequency range permitted to be used for wireless communications lies between 2.4 and 5 GHz. However, within the overall spectrum of frequencies administrated by the FCC there are so called white spaces: frequencies not used by anyone. They are situated between portions used by cable TV and telephone applications. The FCC now proposes to free those unused portions for use in wireless communications. This may open up services employing wireless broadband applications at low cost. White spaces are situated between 500 to 700 MHz. Because of these relatively low frequencies, those radio waves will not be attenuated as much as those currently in use by buildings or the atmosphere, reducing the usual dead communication zones. The challenge will be to develop an appropriate standard. This will take some time so that first products will not be on the market before a year or two from now (2018) [64].

3.8.2 Mobile Hotspots

Up to now hotspots are offered as a service in a fixed location for users passing by or explicitly relying thereon. Considerations now go into the direction of personal mobile hotspots, which could be carried around by individuals. So, even, when there is no traditional hotspot available, people could still connect, if they wanted to.

Mobile hotspots are such devices as routers with uplinks, which can be battery powered, and provide security features, such as encryption and authentication. The devices already on the market use the traditional 802.11a and 802.11g standards, but the fast 802.11n is considered as well. [65].

3.9 Applications

3.9.1 A Small Home WLAN

Figure 3.26 shows a typical example of a small private WLAN, which can be assembled in any household. All that is needed is a router (in this case a Fritzbox) connecting to a private telephone line with DSL capabilities. The two laptops are each equipped with a notebook WLAN adapter. One of the stations has a locally connected printer. The adapters send out signals to search for the emissions of the Fritzbox [66] router. Once the lock on the frequency connection is established, the network is in operation. Note: the notebooks do not communicate directly with one another in this example.

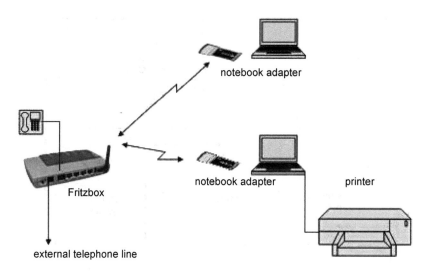

Fig. 3.26. A Small Home WLAN.

3.9.2 Hot Spot Services

Figure 3.27 shows the setup of a hot spot service center acting as intermediary between two public WLANs, the Internet and a company network.

3.9.3 WLAN Solutions for a Hospital

WLAN solutions for hospitals can be extensions of existing LANs. This allows for access to patient and hospital data anywhere in the premises and in all areas—especially there, where traditional cabling is difficult to arrange. In any case personnel becomes more flexible thus increasing their efficiency. New applications, such as store management for pharmaceuticals, mobile ward rounds or IP telephony via WLAN are additional possibilities.

Information access is no longer dependent on specific locations. This enhances cooperation between personnel working in different areas. Unified communications, enhanced security functions (interface to central radio linked alarm systems) and to localization services are possible. Further, applications include local resource management concerning free bed or room capacities or medical equipment. These data can be mapped to a graphical display of the hospital floor plan. Via the radio linked alarm system nurses can place emergency calls or track the movements of disabled patients.

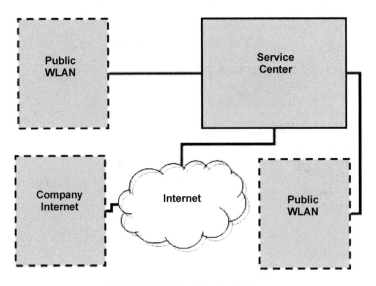

Fig. 3.27. Hot Spot Service Center.

3.8 Checklist

Table 3.3 contains a checklist for WLAN implementation.

Table 3.3. Checklist for WLAN Implementation.

Do you want to set up and operate a WLAN?	When setting up and operating a wireless network additional security aspects compared to a cable network have to be taken into account.
Do you plan a WLAN in a large organization?	In a large organization a WLAN has to be integrated into the total IT strategy including the IT security strategy.
Would you like to set up a small private network?	The technical challenges are basically the same as in an organization; only the formal aspects of IT security management are not relevant.
Does a formal IT security management exist in your company?	IT security management deals with all security aspects, when setting up and running IT installations.
Are the terms of reference for security management documented?	Precondition for a functioning IT security management is a sound documentation.
Are the prevalent security standards taken into account for your security management?	For WLAN ISO 17799 among others is relevant.
Are security criteria documented?	Security is classified according to confidentiality, availability, integrity, etc.

Table 3.3 contd....

...Table 3.3 contd.

After security instructions do participants have to sign a declaration?	Participation in security instructions should be documented in the interests of all concerned.
	Signatures are at the same time legally binding and constitute additional commitment to the organization.
Is compliance to security regulations checked regularly?	Compliance should be checked according to a master plan.
Is WLAN operation documented as part of IT security management?	WLANs can pose significant security risks and thus their operation has to be integrated into the overall IT strategy.
Do rules exist in your company against uncontrolled growth of WLANs?	Quite often WLAN initiatives are started by advanced users; they have to be integrated into the overall management concept.
	Unauthorized utilization of other networks via WLAN access offers an uncontrolled gateway for external attacks.
Are all types of data to be handled by a WLAN specified?	According to security classifications certain company data belong to different security classes.
	Transfer of confidential data should only be possible in encrypted form.
Does a WLAN user guide exist?	A WLAN user guide is part of the IT security documentation.
	Missing guidelines for WLAN utilization lead to uncontrolled growth in an organization.
Do general conditions for authorized WLAN access exist?	The general conditions include training, commitment to the company security standards, usage of hotspots, etc.
	Uncontrolled WLAN access undermines the internal security standards.
Do WLAN guidelines exist for system administrators?	These guidelines determine rules for configuring and operating WLAN components.
Do documents for WLAN training exist?	These documents should also contain security requirements besides technical aspects.
Are administrators and users trained concerning WLAN security?	Because of the increased security risks in connection with WLANs detailed knowledge of the major weak points are of prime importance.
Are WLAN components part of the data backup strategy?	Backup strategy is part of the overall documented IT security strategy and of the operating manuals.
	WLAN components should have special attention within the overall backup strategy.
Are data stored on WLAN components part of the general backup?	WLAN components can carry all relevant configuration data for operation besides the usual business data.
	WLAN components may contain configuration data as well as business data, both of which have to be safeguarded.

Table 3.3 contd....

...Table 3.3 contd.

Does the WLAN security directive undergo regular revisions?	Regular revision of IT security directives—especially for WLANs—is part of the internal security process in an organization.
	State-of-the-art techniques and standards for components vary rapidly requiring regular updates of a directive.
Has a contingency plan been developed?	Security incidents concern attacks on WLANs or LANs via a WLAN to spy data, deposit data, to falsify information, to erase them and to steal WLAN components.
	According to the type of security incident a standard process should be triggered as a specific reaction.
Do contingency plan and error handling constitute part of WLAN instructions?	Different security incidents require adapted reactions to be documented as a process if possible: immediate technical measures, documentation, reporting channels, strategic countermeasures, etc.
Does a proper configuration management exist in your company?	Only the configuration management for WLANs should be allowed to setup the components according to the standards in use in the company while respecting the security directives.
Have the WLAN standards to be employed been decided upon?	The market offers different standards: IEEE 802.11b, 11g for the 2.4 GHz band; 11a, 11h for the 5 GHz band and others. A company should use one single standard.
Has a standard configuration been developed?	To support WLAN components efficiently a standard configuration should have been developed according to security directives.
Is configuration done in wireless fashion?	WLAN components can be configured via cable or in wireless fashion.
	Configurations should not be done wireless if possible to avoid the spying of passphrases and cipher keys.
Is the default password of the manufacturer for routers routinely and immediately replaced?	Routers are generally provided with a password by the manufacturer.
	Included default passwords can be identified without problems in the Internet and therefore should be replaced as a first configuration measure.
Do you employ an individual SSID?	The default settings by the manufacturer of SSIDs should be changed from the onset.
Does your SSID designation enable inference about the user?	Descriptive designations may indicate the area of utilization and therefore the data material.
	The SSID designation must not enable inference about usage to avoid additional incentives for spying.
Is the broadcast of beacon frames of the SSID suppressed?	The broadcast of the SSID indicates to the environment that a WLAN is active.
	The spying on active WLANs with the intention to penetrate is called war driving. This is facilitated by SSID broadcasts.

Table 3.3 contd....

...Table 3.3 contd.

Is your radio link encrypted?	Encryption of radio links is one of the basic measures to prevent attacks from the outside.
	Authentication data and confidential information should be transmitted only in encrypted mode.
Are radio links regularly checked with analytical tools?	Analytical tools allow the detection of successful or attempted non-authorized accesses.
Is access to networks regulated in general?	Access to WLANs and LANs should be regulated by processes conforming to audit standards.
	Unregulated access and links bypass organizational measures of the IT security management.
Do you reduce your transmission power to the minimum necessary?	On the one hand a certain transmission power is required to operate the system; on the other hand this can be a security risk.
	The limitation of transmission power prevents intensive broadcasts and thus the detection of a WLAN operated within the boundaries of a company.
Do you use an Access Control List (ACL) when employing RADIUS servers?	The utilization of the IEEE 801.1x standard allows for additional authentication of users via an Access Control List (ACL); the authentication is routed via a centrally controlled RADIUS server.
Have the locations for the installation of WLAN components been determined?	The geographical location of WLAN components determine the broadcast to the outside and prevents dead spot problems.
	WLAN components can suffer interference from radio waves originating from other technical appliances preventing their operation in the vicinity of them.
Is the ad hoc mode always switched off?	The ad hoc mode allows the setup of a spontaneous WLAN by client to client communication.
	When the ad hoc mode is switched on, unauthorized clients may access the WLAN directly.
Do you switch off the Dynamic Host Configuration Protocol (DHCP), if you use it?	DHCP server automatically allocates IP addresses for the overall network.
	DHCP can be a gateway for attacks by allocating a valid IP address to an intruder in the worst case.
Are the frequency channels in use selected without overlapping?	If the frequency channels are too close to each other this can lead to interferences.
	Interferences can be avoided, when channels are used that are far enough apart at any one time thus having a channel separation as wide as possible.
Is your network checked for dead spots?	Dead spots are created by interferences from other electromagnetic waves, an inadequate geographical arrangement, badly adjusted antennas, insufficient transmission power.

Table 3.3 contd....

...Table 3.3 contd.

	Analytical tools allow the detection of dead spots otherwise being experienced by users as disruptions.
Are WLAN components switched off when not in use for some time?	To avoid unnecessary broadcasts WLAN components not in use should be switched off.
	Broadcasts indicate to external spies and war drivers that a WLAN is in operation.
Do you use WEP encryption technology?	WEP encryption technology uses a symmetrical process providing access points and clients with a common key.
	WEP encryption technology is generally regarded as unsecure, but should be preferred against not encrypting at all.
Do you use WAP encryption technology?	The possibilities of the WAP process are state-of-the-art today. If employed they should be applied to all network components.
	Larger networks should be protected additionally by a RADIUS server with Access Control List administration.
Do you use the IEEE 802.1x mechanisms?	IEEE 802.1x is a framework standard using a RADIUS server.
Do you take care to avoid weak passphrases?	The security of passphrases depends on their length and character combination.
	Weak passphrases are short, consisting only of alphabetical letters and can be found in dictionaries.
Do you use authentication methods relying on reciprocity?	With these methods clients and server exchange information to mutually authenticate themselves (IEEE 802.1x)...
Do you use the pre-shared key method?	With the PSK method for each user cycle a new cipher key between client and access point is generated.
Do you use WEP in contexts with confidential information?	WEP encryption technology uses a symmetrical process providing access points and clients with a common key.
	WEP encryption technology is generally regarded as unsecure, but should be preferred against not encrypting at all.
Do you use WAP2?	WAP2 refers to the standard IEEE 802.11i—an improvement with regard to WAP employing the Advanced Encryption Standard (AES).
Are encryption options checked when buying new components?	When selecting new components it is important to buy those with the highest standard present in the network.
Are cryptographic keys replaced regularly?	To prevent systematic spying, the keys should be changed on a regular basis.
	The key change should proceed according to a fixed time table. Change should take place once per months, at maximum once per quarter.

Table 3.3 contd....

...Table 3.3 contd.

Does your company have a dedicated data protection management?	Data protection management deals with integrity, confidentiality, availability and access security for all data in store in a company.
Are confidential data encrypted on mobile devices?	When using mobile terminals certain data have to be stored locally on the devices.
	Mobile devices run a higher risk of theft and loss. If confidential data have to be stored locally they should be encrypted.
Are internal and external networks defined, to which links may be established?	WLANs can act as a gateway to linked LANs.
Is the WLAN linked to a LAN?	In a linkup between WLAN and LAN the WLAN is the weakest spot against external attacks.
Do you employ security gateways when accessing a LAN from a WLAN?	In many cases the security level of a radio link and its components does not correspond to that of a LAN.
	The high security requirements for the transition from WLAN to LAN can be realized with a security gateway.
Is it possible to block WLAN communication from the LAN?	At the point of transition a total blocking of the WLAN communication should be made a possible option, if required.
Do you operate access points as hotspots?	Hotspots facilitate a wireless and simple access to the Internet.
	When operating access points as hotspots additional security measures are necessary especially when operating a LAN in the same configuration.
Are hotspots connected to a LAN?	In a WLAN to LAN linkage, it is possible to gain access to a LAN via a hotspot.
	Hotspots should be connected to a LAN only through a security gateway.
Is inter client communication permitted?	Inter client communication allows the setup of an ad hoc network without integrated control devices such as access points or routers.
	Inter client communication should generally not be permitted in permanent networks.
Do you have a protection strategy against viruses, worms and Trojan horses?	Most organizations employ special software to scan incoming data against infection.
	The protection strategy against infection should have the same level as that for LANs.
Are your installations protected by firewalls?	A firewall controls data traffic between network segments and the outer world on different communication layers.
	When working with the Internet, firewalls are an indispensable part of the security strategy.

Table 3.3 contd....

...Table 3.3 contd.

Are provisions taken to safeguard the network against technical disturbances?	Technical disturbances can originate from different devices emitting radio signals: microwave ovens, surveillance cameras, etc.
Do you employ WLAN management systems?	WLAN management systems document the configuration, analyses network operation and deals with incidents.
Do you document the results of security checks?	This documentation can be handled by a WLAN management system.
Do you analyze the security relevance of these results?	The analysis results should be documented in a WLAN management system.
Are there regular audits?	Audits consider configuration parameters, access rights, password cycle, and compliance with security directives.
	Audits are necessary because of changing technologies and personal turnover.
Do you employ WLAN analysis tools?	WLAN analysis tools check the installations against unauthorized WLAN operations, find dead spots and evaluate signal quality.
Do you employ penetration tests?	Penetration tests simulate attempts of unauthorized access to the network and thus provide a measure for access security.
Are cross points and authentication server checked regularly?	Checking the functioning of those components is part of a regular system audit.
	Cross points and authentication server are critical components regarding network attacks.
Do you regularly check your clients with respect to their configuration and being up to date?	Especially when security standards have been raised one has to make sure that clients comply.
Do you verify the employment of non-authorized components?	Checking against non-authorized components is part of the security audit.
	Besides using a WLAN analysis tool physical verification of devices should be done as well.
Have measures been defined to protect the network against interferences?	Interferences can be avoided by carefully selecting locations or by switching off or displacing the sources of disturbances.
Do you document disturbances or abnormalities?	All disturbances and security incidents should be documented with the help of a WLAN management tool.
	Analysis of error messages permits the recognition of attack patterns or systematic security loopholes.
Is it possible to access the Internet from the WLAN?	Unprotected Internet access is a favorite target for attacks to a WLAN.
	Internet access has to be protected by a security gateway.
Is hotspot usage regulated?	Once terminals have been equipped with WALN adapters access to freely available hotspots is possible.

Table 3.3 contd....

...Table 3.3 contd.

	The usage of hotspots can be prevented by technical means.
Is the usage of external WLANs restricted?	External WLANs allowed to be accessed should be defined in the security directive.
	Access rules to these networks should be defined amicably with the owner of these networks.
Are appropriate encryption methods employed for data transmission between hotspots and server?	Hotspot operation permits simple wireless Internet access for external users.
	Besides the security provisions given by WLAN standards web authentication and additional protocols for data encryption should be employed.
Is there a master plan after theft of critical components?	Measures have to be taken to prevent that stolen components continue to allow access. All configuration data relating to security have to be replaced within the total network.
Do you have a directive for decommissioning WLAN components?	When decommissioning WLAN components at a minimum all configuration data have to be erased.

4

Mobile Phones

◇◇

4.1 Context

The previous discussions concerning WLAN security have centred around classical terminal devices with their corresponding architecture (laptops, PCs, printer). This part shall cover security risks in connection with the usage of mobile phones in its proper sense.

Of interest for further considerations is the fact that mobile phones pose a risk in serving as gateways for attacks on central applications, when employing either their communication functions or certain services that come with these devices. When using a mobile phone both for telephony and WLAN applications at the same time, additional risk potential arise, which will be discussed in more detail in this section, even though this technology is still in its initial stages. Thus, there is no way to circumvent any aspect of mobile telephony.

Firstly, the architecture of mobile phones will be looked at. On this basis current operating systems will be introduced briefly. These operating systems may include various services relevant to security considerations. So there will be again specific threat scenarios and corresponding countermeasures to protect installations.

And finally a directive tailored to the specific requirements for mobile phones in general will be presented.

4.2 Basic Principles

One has to distinguish between:

- External communication structure and
- Internal device architecture.

The interplay between both may lead to situations where security management becomes important. However, even the pure existence of a mobile phone poses a security risk in itself. To understand this, the basic principles of communication and device architecture and its functioning will be presented in the following.

4.2.1 Communication Structure

The general structure of a mobile network is laid out in Fig. 4.1.

What can be seen is a cellular network in its hierarchical arrangement. The main components are:

- The phone itself
- Base station
- Control unit
- Transmitting station, and
- Switching node.

Provider and end user are connected via the base stations. Base stations can serve several cells. They themselves are managed by the control units. Routing and service switching are carried out by the switching node. Additionally, a number of registers for the administration of subscribers are required. This will be discussed further down.

It is important to note that there is normally no end-to-end connection between mobile phones themselves—in contrast to certain terminal devices in WLANs—since the communication is routed via the network.

The deployment of a mobile phone for WLAN communication is shown in Fig. 4.2.

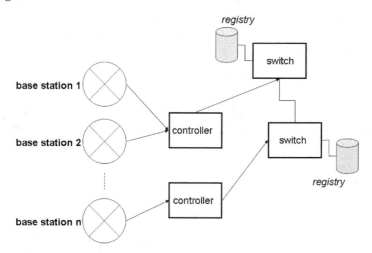

Fig. 4.1. Structure of a Mobile Phone Network.

Fig. 4.2. Mobile Phone in a WLAN.

It is evident that two different protocols are required. The solutions currently available in the market do not necessitate a route via the mobile phone net. Mobile phones equipped for such communications can directly connect with a WLAN via an access point.

4.2.2 Device Architecture

Nowadays, mobile phones have similar capabilities as PCs. Surpassing their initial functionalities for voice communication they are equipped with far more applications. These capabilities carry costs in terms of security risks since the user now possesses additional degrees of freedom. Manufacturers have taken this into account by including in their basic configuration functional modules and a separate security module. The functional modules can be separated into:

- The communications part and
- Local applications.

The main local security module centres around the so called SIM card (Subscriber Identity Module) [67]. The following items are stored on this card:

- Customer ID
- IMSI (International Mobile Subscriber Identity)
- Dial number
- Authentication data.

The physical separation of SIM and device allows the usage of different devices by one and the same end user, since he can carry the SIM card along. The logical connection of the user is therefore to his SIM card and not to the device itself.

Figure 4.3 shows the typical architecture of a mobile phone with its various interfaces.

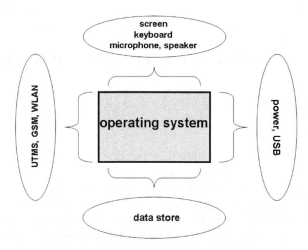

Fig. 4.3. Mobile Phone Architecture.

There are four different types of interfaces:

- User
- Communication
- Storage medium
- Devices.

Each type represents different targets for attack and of relevance to security. The various risk scenarios for these interfaces will be discussed further down.

4.2.3 Smartphones

Smartphones are enhancements of normal mobile phones having led to the integration of comprehensive mobile telephony services and to a glaring multitude of applications (Apps), which was earlier possible only on PCs or laptops. This has only been achieved with the extensive use of the Internet.

As a minimum, the following services and functions are to be found:

- GSM
- UTMS
- GPRS

- HSCSD
- WLAN

as protocols, as well as:

- SMS, MMS
- Emails
- Internet access

for communication. Furthermore, there are other applications such

- GPS
- Office packages
- MP3-Player
- Digital Kameras

and all sorts of Apps administered by the user himself, downloaded and activated. There are no limits to imagination.

Many smartphones use the Android [68] operating system. It was developed by Google. Its program code can be obtained as freeware. Because of this there exists a multitude of modified versions having been adapted by the suppliers of mobile communication devices to suit their products. As a consequence any update has to be organised by those different suppliers. This may lead to a situation, whereby some adapted versions are out of step with regard to the original Android versions. Apps can be downloaded from the Google "Play Store" as well as from other sources. Android phones facilitate memory extensions and can be connected to laptops for example by USB sticks to upload or download data or to synchronise. Synchronisation of addresses and calendars can be made via the Internet.

Figure 4.4 shows the architecture of the Android operating system. At the very bottom is the kernel with the drivers for the basic functions:

- Display
- Camera
- Bluetooth
- USB
- Keyboard
- Wi-Fi
- Audio
- Power, etc.

In the layer above the JAVA runtime libraries are situated. Further up one can find the applications and the applications framework resp. Android keeps these areas flexible so that a developer can integrate his own new applications or replace existing ones.

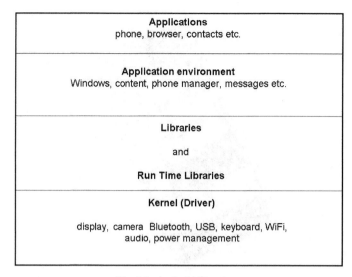

Fig. 4.4. Android Structure.

4.2.4 iPhone

iPhone is a product of the Apple company (Fig. 4.5). Besides its classical telephone functions its main attraction is to be found in the support of medial services as a quasi extended iPod for videos and music. Its operating system iOS [69] is an enhancement of MAC OS X. Its main applications are:

- Webbrowser
- Email program
- Calendar
- Map service
- Notes
- YouTube player
- Pocket calculator
- Weather forecast
- Stock exchange information

and everything else that can be downloaded from the App Store.

iPhones posses a high data storage capacity and at the same time the possibility to connect to central applications via WLAN.

iOS is not only deployed in the iPhone but also on iTouch and iPad. Unfortunately, only Apps bought from the Apple App Store can be installed. For this the user has to register. Memory extension via a separate card is not possible. To exchange data with a PC one has to obtain the program iTunes.

As in other systems the architecture of iOS consists of different layers (Fig. 4.6).

Fig. 4.5. iPhone.

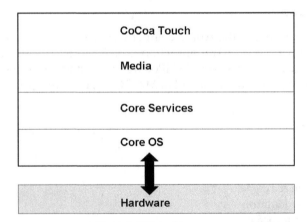

Fig. 4.6. iOS Architecture.

The layers mean:

- Cocoa Touch: adapted from Mac OS X with interfaces and frameworks
- Media: framework for audio, video, etc.
- Core Services: basic system services, memory and data administration
- Core OS: kernel und network functions

4.2.5 BlackBerry

BlackBerry is a very specific type of smartphone brought into the market by the company Research in Motion (RIM) [70]. However, it constitutes a category of its own, which offers its own specific possibilities of usage and, therefore, also of security risks. The manufacturers, being aware of this, provide for appropriate security concepts. These will be discussed separately.

BlackBerrys enhance classical mobile phone functionalities significantly (Fig. 4.7). A BlackBerry is primarily used for the exchange of emails and PIM (Personal Information Management) data. For this purpose it uses a special real time operating software with is own proper communication protocol. Its additional mobile telephone capabilities will be discussed later.

Besides functional and ergonomic ones a BlackBerry offers additional advantages with respect to classical mobile phones. For example, all data are kept synchronously between server and terminal as long as the connection is alive. Its integrated Mobile Data Service (MDS) [71] provides for easy access to internal company databases. Another feature comprises the compression of large quantities of data with the help of the BlackBerry Enterprise Server (BES) [72], which transports these as data streams onto the terminal under acceptable performance. For this special encryption mechanisms are provided for.

At the same time a BlackBerry terminal is capable to join other internal communication systems within a company via the BES and Instant Messaging facility.

Ergonomically the device can be operated with one hand only because of its track wheel and the arrangement of its touch buttons. The more recent versions have dispensed with the track wheel and are equipped with a sure type touchpad.

BlackBerrys are capable to operate within an existing server architecture within an organisation. Thus, a degree of complexity is attained, which asks for corresponding management tools. A special platform with these necessary tools has been developed:

- Push software applications for specified user groups
- Version management
- Specific security modules
- Monitoring possibilities.

Today, all these functions can be executed via one single administrator console—and this under standard WINDOWS© operating systems and firewall protection.

Information concerning the security philosophy of BlackBerry will be discussed later.

Fig. 4.7. BlackBerry (Quelle: Karlis Dambrans – BlackBerry Q10, Flickr).

4.3 Communication Protocols

The operating systems of mobile phones support different communication techniques and protocols. In addition to simple standard functions other functions such as data management and file systems may come along. Cryptographic procedures and access control to protect the device and communications is an integral part of all operating systems. The following popular operating systems will be introduced here:

- GSM [73]
- GPRS [74] and
- UMTS [75].

4.3.1 GMS

Figure 4.8 shows the communication schema for GSM.

The whole network is subdivided into cells. These cells are served by base stations (BTS). They function as interfaces between the provider and the end user at the same time. Furthermore, there are control stations (BSC) managing the resources of the BTS. The BTS are in turn controlled by switching nodes (MSC). The MSC take care of the classic routing including all processing even to fixed networks. In addition there exists a number of registers containing information, without which routing would not function:

- HLR/Home Location Register: information about subscribers (ID, services, etc.)
- VLR/Visitor Location Register: status of the subscriber
- AUC/Authentication Centre: information relevant to authorization validation
- EIR/Equipment Identity Register: list of all approved terminal devices.

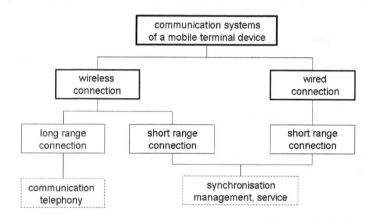

Ref.: BSI

Fig. 4.8. GMS Communication Schema.

4.3.1.1 HSCSD [76]

High-Speed Circuit-Switched Data (HSCSD) is an enhancement of GSM permitting the usage of several GSM radio channels at the same time. This increases the potential data transmission rate.

4.3.2 GPRS

General Packet Radio Service (GPRS) can bundle several radio channels and is most suited for data transmission from the Internet and for sending emails. For this specific services have to be employed—for example i-mode or WAP (s. below).

4.3.3 UMTS

Universal Mobile Telecommunication System (UMTS) represents a new generation of mobile phone operating systems. Because of its optimized transmission mode more elaborate data formats besides text and voice can be transmitted at a high rate: video, Internet, etc. This possibility offers additional services, but also represents new sources for risks.

4.3.3.1 HSDPA [77]

A further enhancement within UMTS is the High Speed Downlink Packet Access (HSDPA) standard, which is most suitable for WLAN applications.

4.4 Services

Apart from classical telephony today one can distinguish between the following additional services:

- General information services
- Other communication services
- Data transmission features

Figure 4.9 illustrates these user services schematically, built into the specific architectures.

The more important features are:

- SMS/EMS/MMS
- WAP and
- i-mode.

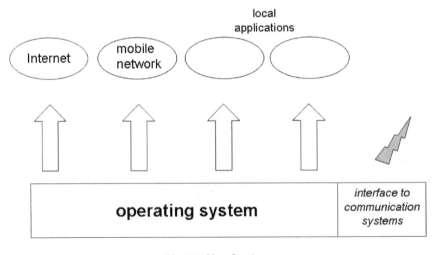

Fig. 4.9. User Services.

4.4.1 SMS/EMS/MMS

Theses abbreviations stand for:

- Short Message Service (SMS) [78]
- Enhanced Message Service (EMS)
- Multimedia Message Service (MMS).

The basic service is SMS. This service allows for transmission of text messages only. EMS and MMS were further developed on its basis.

With EMS, messages exceeding the text limit of SMS (160) as well as containing simple graphics can be transmitted. On the other hand, MMS facilitates sending of photographs and short videos via mobile radio.

A common trait of all three formats is that the messages are not routed directly to the receiver. The addressee gets a notification that a certain message is ready on call. Only once the addressee executes his call will he get the actual message, which is buffered on the server of the provider.

4.4.2 WAP [79]

Wireless Application Protocol (WAP) service secures the transmission of information from the Internet. To be able to use this service the terminal devices have to be equipped with suitable browsers. The WAP architecture corresponds to those of other data networks as can be found in regular client server constellations.

4.4.3 i-mode [80]

i-mode represents yet another Internet access possibility via mobile radio and is therefore in direct competition to WAP. To use its functionality fully terminal devices have to be specially equipped.

4.5 Mobile Phones and WLAN

As described mobile phones communicate via the discussed protocols. The associated billing functions accordingly. The future, however, points into a different direction. Concepts developed for these purposes aim to integrate both mobile phones and WLAN, so that mobile phone access can be realized using WLAN technology, such as access points or hotspots. But this also means additional security risks.

These services are currently handled under FMC (Fixed Mobile Convergence) [81] and UMA (Unlicensed Mobil Access) [82] (s. Fig. 4.10). To implement such services on a large scale a number of technical preconditions have to be created. These concern the terminal device on the one hand and the necessary infrastructure on the other.

Fig. 4.10. Mobile Phones and WLAN Possibilities.

4.5.1 Infrastructure

Concerning the infrastructure these are the areas of interest:

- Convergence of voice networks and WLAN
- Implementation of the relevant networks in organisations
- Provision of the required services by mobile network providers
- Integration of back office systems
- Completely integrated media architecture with all relevant added value services.

Figure 4.11 shows a possible scenario integration of WLAN with UMA.

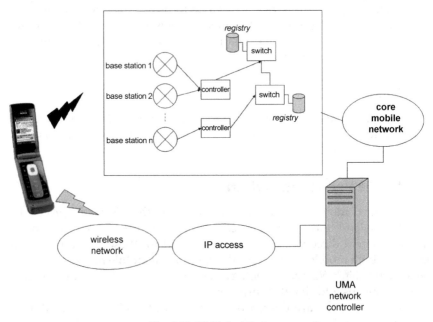

Fig. 4.11. UMA Architecture.

4.5.2 Terminal Devices

Meanwhile the manufacturers of mobile phones are several steps ahead. After some hesitation concerning billing the first phones with WLAN capability have appeared on the market. These have been enhanced accordingly and are also capable of VoIP (s. below). They are usually also equipped with special touchscreens or keyboards. There is still technological improvement required because of high electricity consumption leading to frequent recharges of the accumulators.

4.6 Threats and Protection

4.6.1 General Risk Potentials and Strategic Countermeasures

Security risks emanating from mobile phones surpass largely the classical risks of typical WLAN applications. This is independent from the fact whether they are used for WLAN applications themselves or for telephony only. This means that most dangers are not really dependent on the type of usage but also partly dependent on the nature of the device. For all those reasons countermeasures are necessary that will surpass those mentioned about in a purely WLAN environment. These additional risk potentials will be identified in the following. As usual first measures have to be decided upon on the strategic level. For this purpose the importance of the interplay between strategic, organisational, and technical measures is reminded about (Fig. 4.12).

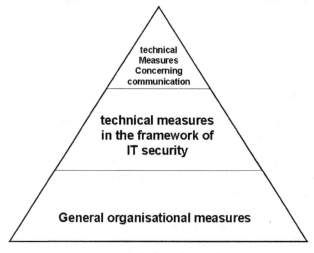

Fig. 4.12. Security Pyramid.

For mobile phones the following general catalogue of risk potential is important (s. Table 4.1):

Besides general measures for all regular mobile phone applications manufacturers provide specific strategies, which have been presented further down with the example of BlackBerry. The following is important for mobile phones:

- Contrary to wired terminals mobile ones are prone to be operated much more often in unsecure environments.
- Besides the technical possibilities to spy on wireless traffic there is a soft spot, which is based on the nature of a mobile phone itself: acoustical eavesdropping.
- Authentication procedures for mobile phones are simple. In most cases a four-digit code suffices to not only gain access to the network under the recorded subscriber identity, but also to locally stored data on the device.
- Besides this all other points of attack known from WLANs and classical wired networks continue to exist.

Table 4.1. Security Risks for Mobile Phones.

Scenario	Target	Type of Attack
Attacker in possession of the device	Applications Hardware Device characteristics Infection	Theft
Attacker does not possess the device	Services DoS Operating system Communication Infrastructure	Manipulation

4.6.1.1 General Organisational Measures

The organisational measures, which are of particular importance to the mobile phone sector, comprise:

- Regulation of the usage of mobile phones.
- Definition of prior procurement criteria with respect to security in addition to those relevant to functionality and costs only.
- Enforcement of security regulations in an organisation, if not already in place (such a strategy should in any case already exist for classical applications in every organisation).
- Development of an authorization concept including a password strategy for mobile phones as well.

- Encryption concepts for data storage and all data traffic.
- Implementation of an alerting process in case of theft or loss of a terminal device including the detection of attempts to gain access illegally.

4.6.1.2 General Technical Measures

General technical measures comprise:

- Monitoring private terminal devices concerning their usage for company functions—especially in company WLANs.
- Clearance procedures for applications on terminal devices.
- Encryption guidelines for communication and data.
- Commitment for physical supervision of mobile terminal devices.
- Directives for data backup.
- Restriction for usage outside the company.
- Inventory of the devices.
- Development of an authentication strategy.
- Encryption of data and storage media.
- Development of an authorisation concept.
- Synchronisation only through secure protocols.
- Integration of the devices into existing operating systems without plugins.
- Monitoring of identification and authentication transactions.
- Deployment of Network Control Software.
- Alerting mechanisms.
- Frequent check of the infrastructure.
- Update of synchronisation programs, communication protocols, and operating systems.

4.6.2 Specific Threat Scenarios Concerning Mobile Phones

The following refers to the risk potentials as outlined in Table 4.1 above. The two scenarios:

- "Attacker in possession of the device" and
- "Attacker does not possess the device"

will be dealt with separately.

Attacks taking place under both conditions have impact on:

- Applications
- Services
- Operating systems
- Communication
- Infrastructure

- Hardware
- Device characteristics
- Infection and DoS

The relevant mobile phone specific situations present themselves as follows:

4.6.2.1 Attacker in Possession of the Device

This is the most dangerous scenario. In this context, following is the summary of the risk analysis:

- Spying on data: personal data, data relevant to security and configuration, business data
- Manipulation of data including deletion: undermining competitiveness of a company by falsifying and sabotage
- Gathering information about company processes: about workflows, about data bank structures, email folders
- Analysis of the operating system, checking security parameters: the acquisition of this information allows deeper penetration into the companies data world
- Manipulation of registry: altering system parameters in a LAN in such a way that usage will be impeded or made impossible
- Introducing viruses, etc.

In addition to those, the following aspects are pertinent:

In the first place an attacker has at his disposal a panoply of technical possibilities to circumvent authentication routines. By this he not only enters the device itself but he then has all the options to penetrate internal and external applications. In any case he has a new launching pad for getting into network applications belonging to a specific organisation. He has surmounted a first obstacle. Using further technical means he can try to conceal his illegal access by manipulation in a way that network administrators or communication officers will not be able to spot his connection as being illegal. By erasing relevant information he is able to alter local records as well such that the original owner of the device, should he get into repossessing it, will initially not be aware of the illegal access (however, once he receives his itemised bill he will discover what happened, but then it will be too late).

Targets of Attack: Services and Operating System

- Analysis of operating system, security configuration
- Manipulation of registry
- Introducing viruses, etc.

Target of Attack: Infrastructure

- Manipulation of the terminal device: illegal use will not be detected immediately
- Destruction of the device
- Duplication and deployment of alien devices: feigning legal access.

These are the further risk potentials specific to mobile phones:

Hardware

The most primitive action is obviously the simple destruction or elimination of the device. Normally this does not present any strategic advantage concerning just a single device. The only benefit of such an action would be to wipe out traces. Otherwise it would not make sense. Much more profitable would be a purposeful hardware manipulation. This could be:

- Deleting of information stored locally
- Manipulation of locally kept data (changing contents)
- Manipulation of applications (uninstalling, introducing malware)
- Setting up a back door for later access to connected systems.

Device Characteristics

Generally speaking, spying on and manipulating device characteristics is a follow up on hardware attacks. Additional dangers result from:

- Exchanging the real device against a dummy with identical characteristics, to be able at a later stage, to pass authentication. After all this, already mentioned possibilities continue to exist.
- Introducing additional storage media, to record the usage. By this a usage profile can be deduced, which could be of interest for various reasons to a spy, such as to uncover additional passwords for external applications. For such an operation to be successful the perpetrator has to come back and arrange for another theft to harvest the additional information or to revert the swap. This could only be possible, if there are systematic security loopholes or completely careless handling.

Infection

Infection means the introduction of malware in its various forms (viruses, worms, Trojan horses). By being in possession of the device the classical venue to circumvent a firewall or a virus scanner within a communication network is no longer required. The attacker is able to implant his vermin in all tranquillity. With the next WLAN access or the next message to an email account it reaches the organisations network and can start to proliferate.

4.6.2.2 Attacker Does Not Possess the Device

This is a situation more difficult for the attacker. But a technically versatile attacker can still wreak havoc employing other means to get access to central applications with the help of an alien device. In this context, following is a summary of the risk analysis:

Target of Attack: Applications

- Spying on data
- Manipulation of data including deletion
- Gathering information about company processes
- Analysis of operating system, security parameters
- Manipulation of registry
- Introducing viruses, etc.

Targets of Attack: Services and Operating System

- Hacking of authentication: by eavesdropping on the radio traffic getting hold of the authentication codes
- Using creeping in methods to get into sessions
- Operating system analysis, uncovering of security parameters
- Manipulation of registry
- Introducing viruses, etc.
- Denial of service: creating massive load on servers that are used for routing messages thus bringing the whole system to a standstill—in other contexts a favourite method to block internet pages.

And these are the additional risk potentials for mobile phones:

By acoustic and electronic eavesdropping an attacker may gather important essential information for getting into the operating system and uncovering vital configuration data. These are for example:

- Authentication parameters
- Access codes for applications
- Man-in-the-middle attacks to join in applications without being uncovered.

Thereafter, by employing certain services further channels open up to create further damage:

Target of Attack: Communication

- Spoofing: all methods to obtain authentication codes, network protocols, system addresses

- Man-in-the-middle camouflage: to sneak between two communicating partners without being noticed and thus obtain all important information necessary to gain access and successively spying
- Protocol attack: changing protocols and mappings to render systems unusable
- Eavesdropping, sniffing.

Target of Attack: Infrastructure

- Theft of the terminal device
- Eavesdropping on data traffic for later inference regarding the infrastructure itself.

Further risk potentials specific to mobile phones:

DoS

Similarly, Internet mobile phone services are prone to the possibilities of Denial-of-Service situations, which can be provoked by flooding a device with data packets or buffer overflows such that a continued usage of the device becomes impossible. Communication has to be interrupted and re-started, which inevitably leads to data loss.

Communication

One has to distinguish between passive and active activities. Passive ones are:

- Eavesdropping
- Sniffing.

The possibilities in connection with eavesdropping have already been discussed. Concerning sniffing one has to distinguish between legal and agreed deployment of a network sniffer for analysis of a LAN or a WLAN against its illegal use. The latter is of interest here. Passive sniffers cannot be traced in the log files of the systems under attack. The following can be intercepted:

- Information about access points
- Data traffic
- Authentication codes.

Sniffers are a favourite tool of war drivers.

The active variants include:

- DoS
- Man-in-the-middle
- Spoofing.

Since mobile phones can also connect to the Internet, spoofing is a relevant threat. Spoofing means pretending a fake identity, for example a webpage supposed to be trustworthy—for example the homepage of a bank. The attacker will try to gather information about the bank account of the user by posing clever questions.

Infrastructure

The only possibilities to influence the infrastructure of a communication network without being in the possession of a legal device are to create disruptions of important communication processes by using one of the already mentioned methods. By gaining information about access codes and sneaking into sessions the attacker has all other possibilities to influence network security and availability.

4.6.3 General Precautionary Measures

4.6.3.1 Data

This is the first rule: store only, what is absolutely necessary. Many data stored and carried around on mobile devices are only of use in the office, where they are stored on other media anyway. The more the information is carried to the outside the higher the possibility that unauthorized persons get knowledge about it and they may destroy or manipulate it.

4.6.3.2 Data Encryption

If it is really necessary to transmit sensitive data via public WLANs or comparable protocols, such data should be encrypted. If possible, business (and private) email queries via mobile phone should be done via SSL connections (Secure Sockets Layer). In this case it is verified, whether a security certificate has been deposited on the message server.

4.6.3.3 Firewalls

Even for mobile phones possibilities exist to implement firewalls. Other protection methods employ the Bluetooth protocol (not to be discussed further here).

4.6.3.4 Encryption on the Device

If possible, locally stored data should be encrypted as well.

4.6.3.5 Backup

Critical information should be saved on a separate device or medium. This claim has been put forward since the early days of IT, but is still not standard behaviour. Because the probabilities for theft and loss are exceedingly higher in this context than those concerning traditional configurations this demand is more important than ever. Even after theft, organisations have to be able to continue to work normally.

4.6.3.6 The e-commerce Risk

Possibilities offered by the Internet or the usage of i-mode or WAP open up completely new risk potentials. These are similar to those mentioned earlier in connection with Internet banking: an attacker could use technical means to spy on bank data und misuse them later. Nothing at all should go without encryption.

4.7 Special Case Blackberry

The manufacturers of Blackberry concede that they do not provide additional security features other than those already discussed and above the usual standard concerning SMS and MMS traffic. SMS and MMS are not encrypted on BlackBerry. The manufacturers propose a number of other measures to increase security. These comprise:

- A special policy in organisations regulating external connection with BlackBerrys. This is a typical task for a directive appropriate for all kinds of handhelds.
- "Confirm on send": requesting the user to confirm that he really wants to transmit the message before actually sending it
- Neutralize the forwarding function (prevents the dissimulation of malware)
- Neutralize the possibility to communicate via non-encrypted messages between users without passing through a secure server.

There are additional possibilities to eliminate non-secure messages from BlackBerrys by routing them only through a suitable BlackBerry server environment. In this way there exist the following configuration options:

- PIN messaging off
- SMS messaging off
- MMS messaging off.

This of course will take away some of the fun from the game, but excludes important security potentials at the same time.

Another possibility is the deployment of S/MIME [83] technology for emails. This technology is based on a general security philosophy relevant for any kind of communication and is not restricted to wireless networks only. Therefore, it will not be discussed any further in detail here. S/MIME packages are on offer by the BlackBerry manufacturers and can be acquired separately. S/MIME supports among others:

- Certificate validation
- Cipher key synchronisation
- Encryption
- Support of smart cards [84].

The manufacturers of BlackBerry, RIM (Research in Motion), have developed a whole panoply of security mechanisms, which come with the system. These comprise:

- A holistic security concept for organisations
- Special encryption algorithms
- Exceptional security measures for WLAN message transmissions
- Exceptional protection for the component architecture
- Local user authentication
- Facilities for device control
- Special measures after theft or loss.

In this book some exemplary aspects will be followed up.

4.7.1 The Overall Security Concept

The BlackBerry solution is based on the Symmetric Key Cryptography [85]. This method ensures confidentiality, integrity, and authenticity.

Confidentiality is assured by encryption based on a secret cipher key, which is known only by the intended receiver.

Integrity is achieved by a random process. The BlackBerry device sends data together with one or more message keys generated randomly to avoid decryption by third parties. Only the BlackBerry Enterprise Server and the device itself has knowledge of the master key and the format used.

Authenticity is guaranteed through authentication of the device by the BlackBerry Enterprise Server using this master key.

Maximum data protection is achieved by:

- Encryption of all data traffic between the BlackBerry Enterprise Server and the terminal device
- Encryption of data traffic between the message server and the email system of the user

- Encryption of data on the terminal device itself
- Encryption of configuration parameters
- Local user authentication on the device by a smart card with password or passphrase.

4.7.2 Security in WLANs

The BlackBerry security concept attempts to harmonize advanced security approaches with existing network technology. The aim is that the end user can send and receive messages without problems in a secure environment without being bound to a particular work place. The interplay with a BlackBerry Enterprise Server functions along the lines outlined in Fig. 4.13

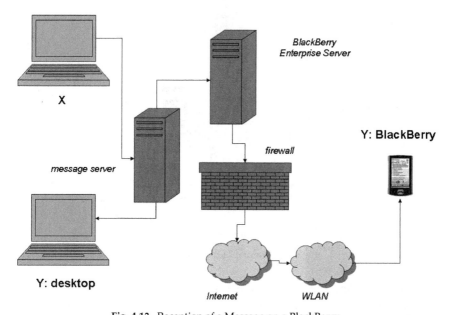

Fig. 4.13. Reception of a Message on a BlackBerry.

X sends a message to Y from his desktop. X and Y work in the same company.

The message server receives the email and informs the BlackBerry Enterprise Server about the reception.

The message server delivers the message to the desktop of Y.

The BlackBerry Enterprise server takes the message from the message server.

The BlackBerry Enterprise server queries the message server, if the forwarding function for Y's BlackBerry mobile has been enabled.

The BlackBerry Enterprise server packs and encrypts the message.

The BlackBerry Enterprise server puts the message into the sending queue.

The WLAN routes and delivers the encrypted message to Y's BlackBerry.

Y's BlackBerry receives the encrypted message, decrypts, and displays it.

The reverse process—sending a message from a BlackBerry inward—goes along similar lines (s. Fig. 4.14).

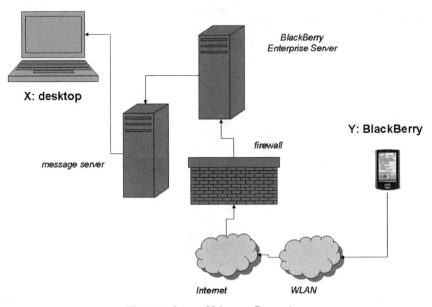

Fig. 4.14. Internal Message Processing.

Y sends a message from his BlackBerry via the WLAN.
The message is transmitted via the Internet.
The message has to pass the firewall of the central network.
The message arrives on the BlackBerry Enterprise server.
From there it is forwarded to the message server.
Finally it appears on the X's desktop.

4.7.3 Theft or Loss of BlackBerry

There are a number of counter measures, which are suited to trigger certain security mechanisms even after a BlackBerry has got into the possession of unauthorized third parties. These include:

- Protection of applications by passwords
- Detection of accesses from logfiles
- Restriction of download possibilities
- Locking of the device
- Remote disabling and deletion of certain data as soon as an access attempt has been started from the stolen device
- Deletion of master keys
- Decoupling of the smart card from the device.

4.7.4 Management Platforms for BlackBerry

To save costs and to increase security, management platforms have been developed for large-sized BlackBerry applications. These management systems fulfil the following requirements:

- Pushing of specific software applications to certain users or user groups
- Administration of versions, upgrades and group memberships
- Detect, notify and delete non-authorized applications on individual BlackBerrys
- Support and administrate thousands of BlackBerry user and a multitude of BlackBerry enterprise servers
- User logging, authentication, and analysis.

These functionalities are the basis for status reports dealing with the following queries:

- What does the BlackBerry infrastructure look like?
- Which applications are in current use?
- What kind of software is implemented on the terminal devices?
- What activities are currently going on in these devices?
- How many devices of which type are in use?

4.8 VoIP

Voice over IP (VoIP) [86] facilitates telephony using the Internet. Using this technology classical telephone infrastructure can be avoided. Various terminal devices can be employed to this end. After the development of smartphones the same possibilities now exist for the mobile phone sector. For this purpose smartphones utilise the WLAN protocol to gain access to the Internet via its access points.

4.8.1 Additional Security Aspects Concerning VoIP

The more important ones are:

- Spoofing (s. above)
- DoS (s. above)
- SPIT (SPAM over Internet Telephony) [87]
- Vishing (faking false hotlines to obtain confidential user data).

4.9 Security Check

The most serious dangers for security mostly occur in locations extern to an organisation, when the user is travelling:

- In public transport
- In hotels
- In the premises of business partners.

Because of time pressure, overwork, and reduced concentration, chances of a device getting lost or stolen rise. But even within the buildings of an organisation there are certain risks. To face the overall security potential a regular risk analysis at fixed time intervals should be carried out. For this purpose, the following Table 4.2 may be helpful:

This table should be completed from the heads of departments or groups in a company. The result of the analysis of these entries gives a qualitative overall picture of the actual threat potential within an organisation.

Table 4.2. Risk Analysis.

Question/Starting Position	Weight
Some employees work remotely from headquarters.	
Many employees travel regularly and need remote access to electronic communication media.	
Many employees need mobile phones to accomplish their work.	
Many employees possess company mobile phones.	
There is no clear policy concerning qualification for company mobile phones.	
There is no clear accountability for the distribution of mobile phones.	
There are no explicit rules regarding cost accountability for mobile phones.	
Against effective directives users still connect to central systems.	
End users create and download confidential data.	
Company culture as a whole is not very restrictive.	

Weights may be assigned as usual:

- 1 for totally untrue
- 2 for probably untrue
- 3 possibly
- 4 probably true
- 5 absolutely true.

The results should be collected and evaluated by the security organism in the company. Departments obtaining a high hit rate should attract special attention. In certain cases measures should be taken to reduce the evaluated risk. To maintain sustainability such an audit should be carried out on a regular basis.

4.10 Directive

The following directive is a tentative example and can be adapted to the needs of a particular company or organisation:

Directive for the Usage of Mobile Phones

The company provides mobile phones for a limited number of users. To obtain a user license a special application has to be filed (mobile phone application form). After assessment a license may be granted.

In special cases superiors may apply for mobile phones for their staff members. This is possible, if compelling business requirements can justify this. The application can be filed informally to the IT security officer.

The management of the communications department is responsible for the selection of compatible mobile phones and for organizing their support. Decisions concerning user authorization do not fall within its field of competence, which is also true for the provision of necessary financial means for the procurement of individual devices. Costs for purchase and maintenance of mobile phone usage have to be borne by the different departments themselves.

Mobile Phones, which belong to the company, may only be used for purposes in connection with the company's business. Any personal usage is excluded. Expenses that occur through personal usage by the user have to be borne by the user himself. Expenses that occur for the company through such personal usage will be billed by the company to the user.

It is prohibited to install non-authorized software on company mobile phones. It is equally prohibited to download additional software or services (for example ringtones) on mobile phones.

It is forbidden to dock non-authorized mobile phones onto other company devices, such as computers, laptops, servers or networks, to connect with them or synchronize with them without prior written permission.

Employees, who obtain a usage license, are responsible for the security of these devices. The devices have to be carried along permanently throughout a business trip. Stolen or lost mobile phones will have to be replaced by the user. The devices remain property of the company.

Sensitive and confidential information may not be stored on mobile phones. In case of loss or theft the IT security officer has to be informed without delay to trigger the necessary steps for remote deletion of contact, calendar and configuration data.

Non-compliance

Any non-compliance incident against this directive has to be notified to the IT security officer. Non-compliance may lead to disciplinary sanctions up to the dissolution of the employment contract. This is independent from other legal actions.

Acknowledgement of the Instruction

This directive should be part of comprehensive employee security instructions. At the end, the following agreement can be signed:

Acknowledgement of the Mobile Phone Directive

"Please read the present mobile phone directive and countersign it at the bottom of the document. One copy with your signature will be kept by the IT security officer.

With your signature you acknowledge:

1. I have received the mobile phone directive, understood its meaning and agree with it.
2. I confirm that I will use mobile phones handed out to me by my employer exclusively for the company's business activities.
3. I agree that I shall carry all costs, which may occur for the company as a consequence of my private use.
4. I will not connect mobile phones to computers, laptops, servers, systems or networks, which have not been cleared for this.
5. I will not store confidential and security relevant data on mobile phones.
6. I understand that I am responsible for the security and replacement of the device after loss. The device remains the property of the company.
7. I understand that non-compliance regarding this directive can induce legal consequences.

Name:
Signature:
Department:
Date:

A directive (with or without signature) has only limited power to prevent havoc. Important is the general attitude of all parties concerned and the necessary discipline.

4.11 Checklist

Table 4.3 summarizes all critical checks for the deployment of mobile phones:

Table 4.3. Checklist Mobile Phones.

Are the responsibilities for security clearly regulated (strategically, organisationally, technically)?	The organisational security procedures comprise general organisational measures, technical measures concerning IT security, technical measures regarding communications.
	Unresolved responsibilities endanger regular operations.
Do the security guidelines contain a catalogue of countermeasures in case of security incidents?	Depending on the type of incident different countermeasures take effect.
	A classification of countermeasures is necessary.
Has the usage of mobile phones been regulated?	Besides the already existing security risks in WLANs mobile phones constitute completely new types of risks.
	High mobility and additional communication potentials increase potential security risks significantly.
Is a general security check in place for mobile phones?	To confront the overall risk potential risk analyses should be carried out on a regular basis.
	A security check documents the distribution and usage of mobile terminals in an organisation.
Are private mobile phones in use?	For reasons of cost savings or for freelancers private phones will be admitted occasionally.
	When these phones can be used for private purposes control mechanisms will fail.
Are company owned mobile phones being deployed?	Generally companies provide their own mobile phones to their employees.
Are mobile phones only used for voice traffic?	Voice is the normal mode of mobile phone usage.

Table 4.3 contd....

...Table 4.3 contd.

Are mobile phones used to transmit data as well?	There are various possibilities to transfer data by mobile phones.
Are mobile phones used in WLANs?	Specially equipped mobile phones permit a connection to WLANs (standard upgrade HSDPA).
	When transmitting confidential data special security precautions must be respected.
Are mobile phones used for the company's business outside the organisation?	Normally mobile phones will be used outside the company.
	The use should be regulated on the basis of directives.
Are remote accesses to central applications frequently necessary?	Mobile employees need remote access.
	A separate security concept should be developed for regular remote accesses to applications.
Do many employees work with mobile phones from outside?	In some departments (field service, external assembly) all employees are equipped with mobile phones.
	A separate security concept should be developed for regular remote accesses to applications.
Has the deployment of mobile phones outside the organisation been regulated restrictively?	The usage should be regulated through appropriate directives.
	Usage of company mobile phones should not be left to the discretion of the user.
Is the usage of mobile phone permitted in hotels?	Hotels represent "unsecure areas" concerning communication security.
	Protective attitudes against eavesdropping are important.
Are mobile phones used on the premises of business partners?	The premises of business partners are "unsecure areas" concerning communication security.
	Protective attitudes against eavesdropping are important.
Have security directives regarding WLAN usage been documented?	Security directives constitute a separate area of a companies IT security strategy.
	Security aspects with regard to WLANs should be documented separately.
Are rules regarding the distribution and authorization for mobile phones in place?	Such rules should be part of directives.
	Non existing rules will lead to uncontrolled growth.
Have selection criteria been defined for the purchase of these terminals?	A catalogue of criteria is useful to sound the market. Special attention should be given to security aspects.

Table 4.3 contd....

...Table 4.3 contd.

	The purchase department should receive guidelines with appropriate criteria.
Are authentication procedures employed?	Authentication procedures are the main prerequisite for WLAN security.
	Without authentication procedures no WLAN should be operated.
Are applications protected by password?	Individual applications can be protected by passwords.
	Protection by passwords are a matter of course.
Is a password change strategy in place?	The change strategy comprises the rate of change as well as the password format.
	Change strategy should not be left to the discretion of the users.
Has a password change cycle been defined?	For the change cycle a time schedule should be communicated. Change should take place either monthly or at most quarterly.
	Passwords should be supplied with an expiry date.
Are complex password structures required?	The security of passphrases depends strongly on their length and character combination.
	Simple passwords can be guessed by an attacker easily.
Do alert procedures exist in case of security incidents?	Every organisation should have a suitable alert process in place.
	Without any alert process a timely intervention is not guaranteed.
Do control mechanisms exist concerning deployment und usage of private terminals in the company?	Usage of private terminals should be subject to mandatory control mechanisms.
	In exceptional cases private devices may go online under defined conditions.
Are terminal devices inventoried?	The IT security officer together with IT management is responsible for documenting the existing infrastructure.
Will synchronisation programs and communication protocols be updated regularly?	These software components should be kept at the most recent technical level to warrant continued compatibility and support by the manufacturer.
Is logging analysis carried out regularly to control access?	An appropriate management platform permits user logging, authentication and analysis.
	Logfiles register also access attempts by non-authorized persons.
Is the infrastructure checked regularly?	This comprises physical inspection as well as checks on all relevant communication protocols and log files.

Table 4.3 contd....

...Table 4.3 contd.

	Checking the records should follow a regular schedule.
Are operating systems and hardware being brought up to date on a regular basis?	Updates and Upgrades provide the latest standards in security.
	Falling behind in software and hardware levels might endanger warranty.
Are counter measures against eavesdropping in place?	Only by an efficient combination of organisational and technical measures can the overall risk be reduced.
	Eavesdropping for some length of time may reveal even encrypted passwords and data.
Are itemised bills analysed regularly?	Itemized bills uncover non-authorized usage.
	Itemized bills should be ordered as a matter of course and analysed.
Is one and the same device in use by several users?	Some companies allow this mode of operation by changing users for cost reasons.
	Changing users complicates the enforcement of control mechanisms.
Are users provided with SIM cards only?	Some users could be equipped with a SIM card instead of a dedicated mobile phone.
	SIM cards could get lost more easily. Besides there is no control about the terminal device on which they will be installed.
Is a SIM card management system in place?	SIM cards should be inventoried.
	Only an inventory allows the correct allocation to a user.
Is GSM employed?	GSM is the most popular mobile phone standard.
Is UMTS employed?	UMTS provides additional functions with respect to GMS (MMS for example).
Do mobile phones access e-commerce applications?	By using different protocols (Internet, WLAN) e-commerce transactions can be executed.
	e-commerce transactions have to be protected against spying (encryption).
Do local applications run on mobile phones?	Office and other applications can be installed or activated.
	Local applications could present an additional source for viruses.
Do clearing procedures exist for applications on terminals?	Only software for company needs and certified for this purpose should be allowed to go on stream for mobile phones.
	Private applications should have no place on such devices.

Table 4.3 contd....

...Table 4.3 contd.

Are business data stored locally?	Some applications required local data storage.
	Local business data should be kept to a minimum.
Are only absolutely necessary data kept?	Some applications require local data storage.
	Minimum and maximum of permitted data should be fixed centrally.
Are memory extensions permitted?	Memory extensions facilitate uploads of spread sheets and other data.
	Memory extensions also facilitate uploads of unwanted applications.
Are central data bases accessed?	It is possible to tap confidential information from central databases by remote access.
	Remote access of central data bases has to be controlled by special security provisions (encryption).
Are regular backups carried out for local data?	Backups become important, once a device has been lost or stolen.
	Without backup lost information has to be reconstituted tediously by hand.
Are data encrypted?	Encryption of data and storage media is a basic part of security strategy.
	Non-encrypted data can be spied on without effort.
Is SMS traffic permitted?	Via SMS short messages (160 characters) can be transmitted.
	SMS messages should not contain confidential information.
Is EMS traffic permitted?	EMS facilitates the transmission of longer messages by stringing together several SMS.
	EMS messages should not contain confidential information.
Is MMS traffic permitted?	MMS can be used to distribute photos or videos.
	MMS messages may contain viruses in their appendices.
Is i-mode being deployed?	i-mode allows Internet access with mobile phones.
	Internet access should be protected separately.
Is WAP being deployed?	WAP abbreviates Wireless Application Protocol. This service allows the transmission of information from the Internet.
	Internet access should be protected separately.

Table 4.3 contd....

...Table 4.3 contd.

Is Internet access via mobile phones permitted?	The possibilities concerning Internet usage should be regulated by directives.
	Internet access should be protected separately.
Is UMA technology being deployed?	UMA facilitates WLAN access via mobile phones.
	UMA technology integrates voice and WLAN.
Does the organisation already integrate voice and WLAN?	Integration of voice and WLAN means integration of voice and data.
	The combination of voice and data presents extraordinary challenges concerning security measures.
Is VoIP for WLAN being deployed?	WLAN also facilitates the implementation of VoIP strategies.
	This will permit connections via access points or hotspots.
Have email functions been defined for handhelds?	Any account structures should be specified beforehand.
	Email traffic should be reduced to the necessary.
Are central applications protected by firewalls?	Firewalls are indispensable preconditions for the secure operation of any network.
	Without firewalls no communication system should go live.
Are the access paths to central systems protected by virus scanners?	Virus scanner should be standard.
	Without virus scanners no communication system should go live.
Are BlackBerrys permitted for mobile phone services?	BlackBerrys possess an independent security architecture.
Have rules been developed for the employment of BlackBerrys?	Their usage should be regulated by directives.
	BlackBerrys require their own security provisions.
Is it possible to circumvent the server without encryption?	There is a theoretical possibility to transmit non-encrypted messages.
	Transmission of non-encrypted messages should be prevented.
Is the option "confirm on send" activated?	This features requires an additional confirmation to prevent accidental transmission of non-encrypted or confidential information to the wrong addressee.
	The option "confirm on send" should be used if possible.
Has the forwarding option been deactivated?	De-activation of the forwarding option prevents the proliferation of viruses.

Table 4.3 contd....

...Table 4.3 contd.

	De-activation of forwarding should be considered if useful in certain cases.
Has the function "PIN messaging" been de-activated?	PIN messaging facilitates message traffic to bypass the BlackBerry Enterprise Server.
	PIN messaging should be set to "off".
Has SMS been de-activated?	For BlackBerrys SMS functions can be switched off.
	De-activation of SMS should be considered if useful in certain cases.
Has MMS been de-activated?	For BlackBerrys, MMS functions can be switched off.
	De-activation of MMS should be considered if useful in certain cases.
Has S/MIME been deployed?	S/MIME is a security package that provides additional encryption protection.
	For highly sensitive applications this additional investment should be considered.
Is the use of the digital camera permitted?	Many mobile phones have a digital camera integrated.
	The digital camera should only be used when documenting company specific information.
Are MP3 players permitted?	Many mobile phones have MP3 players integrated.
	MP3 functions facilitate the download of files which may contain viruses.

5

Bluetooth

◇◇◇

5.1 Introduction

In the preceding chapters the basics and the security aspects in the wireless domain were mostly related to WLAN standards. Although the WLAN is the most important standard in this context, it is by no means the only one in wireless communication. In the following, a communication standard for small distances will be introduced: Bluetooth.

Initially the technical basics, such as protocols and system topologies will be dealt with. Then procedures concerning implementation and configuration will be discussed. After this the security aspects come into play. These security aspects deal with existing security features, risk potentials and possible countermeasures. In the end current developments and the future of this technology will be touched upon.

5.2 Technical Basics

The technical basics of Bluetooth [88] comprise:

- Protocols and
- System topologies.

Protocols are subject to continued review as is the case with all other protocols in wireless context in general. Its development will be traced back in the following section. Network construction, thus, depends on the state of these advancements.

5.2.1 Protocols

In the year 1998, the Bluetooth Special Interest Group (SIG) was founded with the intention to develop an authoritative communication standard for very short distances. One year later the standard 1.0a was proposed, and already at the end of that year version 1.0b was published (the features of the relevant versions will be discussed later on). In the beginning of 2001 version 1.1 was introduced as the first usable market standard.

Following was Bluetooth 2.0 (2004) and 2.1+EDR (Enhanced Data Rate) (2007) [89]. Version 3.0+HS (High Speed) [90] followed and provided an additional channel for WLAN use (2009). 3.0 was available for Enhanced Data Rate (EDR) as well.

The standard 4.0, also called Bluetooth Low Energy [91], the specifications for which were released in 2009, and which was brought to market in 2010, is not downwards compatible with its predecessors. Its main advantages are:

* Reduction of energy usage (important for smartphones)
* Connection establishment in less than 5 milliseconds
* Range up to 100 m.

New improved versions with special emphasis on security were released in 2013 (4.1) and 2014 (4.2 Smart).

Version 5.0 will be discussed further down in a special section.

5.2.1.1 Structure

Figure 5.1 shows the structure of the Bluetooth protocol [92]:

Besides the usual communication elements special attention should be given to the Link Manager Protocol (LMP) featuring the security checks. The remaining abbreviations stand for:

* IP   :   Internet Protocol
* L2CAP   :   Logical Link Control and Adaptation Protocol
* OBEX   :   Object Exchange Protocol
* PPP   :   Point-to-Point Protocol
* RFCOMM   :   Radio Frequency Communications
* SDP   :   Service Discovery Protocol
* TCS   :   Telephony Control Specification
* Cal   :   Calendar
* WAE   :   Wireless Application Environment
* WAP   :   Wireless Application Protocol

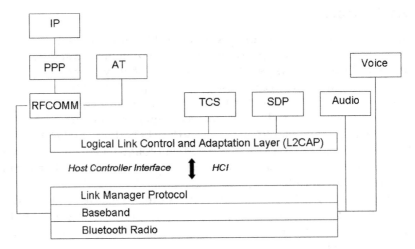

Fig. 5.1. Bluetooth Protocol.

5.2.1.2 Common Protocols and Performance

Table 5.1 shows the Bluetooth versions together with their capability characteristics:

Table 5.1. Bluetooth Versions.

Version	Transmission Rate
1.0	732,2 KBit/s
1.1	732,2 KBit/s
1.2	732,2 KBit/s
2.0 + EDR	2,1 MBit/s
2.1 + EDR	2,1 MBit/s
3.0	24 Mbit/s

In addition to these characteristics one distinguishes three performance classes (Table 5.2):

Table 5.2. Bluetooth Performance Classes.

Class	Power in mW	Range in m
1	100	100
2	2,5	10
3	1	1

5.2.1.3 Features

Bluetooth protocols are suited for data transmission over short distances for:

- Mobile phones
- Mouse
- Laptops
- Printer
- Digital cameras
- Video cameras
- Web pads
- Loud speakers
- Television sets
- Earphones
- Hands free kits
- Other similar devices.

(Meanwhile Bluetooth has entered the entertainment and toy realm; however, these applications will not be followed up here.)

A special feature is the possibility to define certain usage profiles [93] for data exchange. These profiles relate to the communicating devices to be employed. Table 5.3 lists a selection of such profiles:

Table 5.3. Bluetooth Profiles.

Abbreviation	Designation	Meaning
A2DP	Advanced Audio Distribution Profile	Streaming of audio data
AVRCP	Audio Video Remote Control Profile	Remote control for audio/video
BIP	Basic Imaging Profile	Transmission of image data
BPP	Basic Printing Profile	Printing
CIP	Common ISDN Access Profile	ISDN connection via CAPI
CTP	Cordless Telephony Profile	
DIP	Device ID Profile	
DUN	Dial-up Networking Profile	Internet dial-up connection
ESDP	Extended Service Discovery Profile	Enhanced service recognition
FAX, FAXP	FAX Profile	
OBEX-FTP	File Transfer Profile Object exchange	

Table 5.3 contd....

...Table 5.3 contd.

Abbreviation	Designation	Meaning
GAP	Generic Access Profile	Access management, basic profile
GAVDP	Generic AV Distribution Profile	Transmission of audio/ video data
GOEP	Generic Object Exchange Profile	Object exchange
HCRP	Hardcopy Cable Replacement Profile	Print application
HDP	Health Device Profile	Secure connection between medical devices
HFP	Hands-free Profile	Cordless telephony in cars
HID	Human Interface Device Profile	Data entry (from USB specification)
HSP	Headset Profile	Voice output per headset
ICP, INTP	Intercom Profile	Radiotelephony
LAP	LAN Access Profile (nur Version < 1.2)	PPP network connection (more recent PAN)
MAP	Message Access Profile	Message transmission between devices
OBEX	Object Exchange	Generic data transmission between two devices
OPP	Object Push Profile	Transmission of separate files (images, music, business cards, dates)
PAN	Personal Area Networking Profile	Network connections
PBA, PBAP	Phonebook Access Profile	Access to telephone directory (read only)
RS-232	Serial Port Profile	Virtual serial interface
SAP, SIM, rSAP	SIM Access Profile	Access to SIMcard (rSAP remote)
SCO	Synchronous Connection-Oriented link	Access to microphone and earphone of a headset
SDAP	Service Discovery Application Profile	Identification of existing profiles
SPP	Serial Port Profile	Serial data transmission
SYNCH, SYNC	Synchronisation Profile	Data comparison
VDP	Video Distribution Profile	Transmission of video data
WAPB	Wireless Application Protocol Bearer Carrying WAP point-to-point over Bluetooth	

5.3 System Topology

Bluetooth uses a frequency range between 2,400 and 2,480 MHz. On top of this a radio connection to fixed line telephony is possible. Altogether two different data channels are provided:

- Synchronous (SCO) for voice
- Asynchronous (ACL) for all other types of data.

5.3.1 Topology

The network, within which Bluetooth devices communicate, is called a Piconet. Such a Piconet [94] is assembled by the participating devices themselves. The number of devices that can be assembled in such a network amounts theoretically to 255. However, of those only eight can be active at any given time. To function one device has to be designated as master. The other seven are called slaves. One and the same Bluetooth device can also be connected to several different Piconets at the same time—as long as it does occupy the master role (Fig. 5.2 and 5.3).

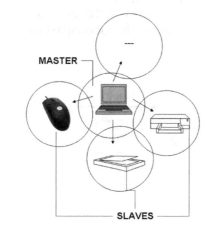

Fig. 5.2. Piconet.

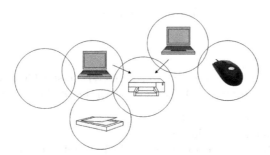

Fig. 5.3. Piconet Overlap.

5.3.2 Connection Set-Up

Every Bluetooth device has a device address with a length of 48 bits. A Bluetooth device continually polls its environment (Inquiry) to find out, whether another Bluetooth device is within its reach and wants to communicate. Once devices have been recognized a paging request is initialized to set up a distinctive connection. The paging device then functions as master and discloses its address. Within a Piconet point-to-point connections as well as 1-to-n connections can be initialized.

5.4 Connecting to the Network

Bluetooth devices possess a device address of 48 bit length (BDA: Bluetooth Device Address). Connection can only take place once the device is active (Fig. 5.4). If this is the case, the device address is broadcast every two seconds. At the same time the device is searching for other devices within its transmission range every 5.6 seconds (Inquiry). For this to happen, the search function has to be activated on the device. The initiating device will be master after successful connection.

The master device then sends per "Paging" its address and its timing cycle to the slave(s). Only after this the connection process will be completed.

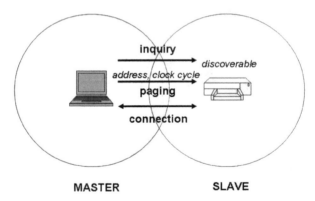

Fig. 5.4. Connection Setup.

5.5 Version 5.0

Here are the most important improvements with respect to its predecessors delivered by Bluetooth V 5.0 [95], released in December 2016:

- Greater range (up to 200 m)
- Higher data rate (2 Mbit/s)
- Energy efficiency (fort the version BLE: Bluetooth Low Energy).

At the same time Bluetooth V 5.0 defines new performance classes (s. Table 5.4).

However, if the error correction mode is used, the data rate may drop to 500–125 Kbits/s. Energy saving is achieved either by increasing time intervals between paging calls or by combining user data transmission with paging calls without initially establishing a permanent connection to another device.

Table 5.4. Performance Classes in V 5.0.

Class	Power in mW
1	100
1.5	10
2	2,5
3	1

5.6 Security Aspects

As is the case with other communication protocols Bluetooth is also prone to attacks from outside. These risks are in part identical to those known for WLANs, partially specific, because they are related to Bluetooth technology. In the following those security mechanisms provided by Bluetooth as a standard will be presented first. Thereafter, the general and specific risk potentials will be identified before discussing the appropriate countermeasures to neutralize such risks.

5.6.1 Instruments

Bluetooth uses various system specific security relevant adjustments and possibilities. These contain:

- Security operations modes
- Cryptographic mechanisms
- Authentication
- Encryption.

5.6.1.1 Security Operations Modes [96]

Bluetooth provides different operations modes. They stand for different levels of security. They are effectively:

- Mode 1 (Non Secure): no special security provisions, no authentication required
- Mode 2 (Service Level Security): security mechanisms at the service level

- Mode 3 (Link Level Security): security mechanisms at the link level—cryptographically (authentication) and/or data encryption.

5.6.1.2 Cryptographic Mechanisms

The basis for the cryptographic method are link cipher keys in connection with the so called pairing method [97] between two devices. This cipher key (length: 128 bits) constitutes itself from a combination of the device addresses and a random number for each device. The generated random numbers will be transferred to the other device respectively. To render this transfer securely an initialisation cipher key is required to be constituted from the following elements (Fig. 5.5):

- Another random number
- Address of one of the involved devices
- PIN.

The PIN has to be identical for both devices (length: up to 16 Bytes).

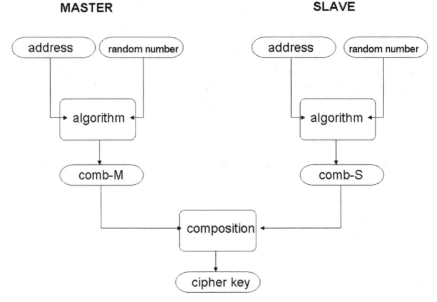

Fig. 5.5. Key Construction.

5.6.1.3 Authentication

Authentication works (initialized from one side) from device to device (point to point). The following automatism will be executed (Fig. 5.6):

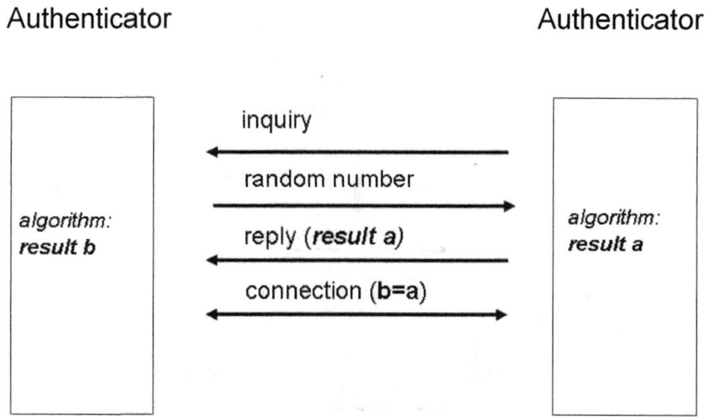

Authenticator   Authenticator

Fig. 5.6. Authentication.

- Authentifier sends random number to authenticator
- Authenticator calculates a reply from the random number, the combination cipher key and its own address (32 bits)
- Authenticator sends reply to authentifier
- Authentifier executes the identical calculation. If both results are found to match, the desired link will be established.

5.6.1.4 Encryption

Encryption (Fig. 5.7) can only take place after authentication and establishment of a stable link. For this to happen another cipher key has to be agreed upon. It is constituted from:

- The combination cipher key
- An offset and
- A random number.

For encryption two modes of operations are offered:

- Point to point or
- Point to multi-point (master to several slaves in the piconet).

Encryption is provided for data transfer only by using a stream cipher made of:

- Device address
- Encryption cipher key
- Timing cycle of the master.

Encryption on the terminal devices has to be done separately.

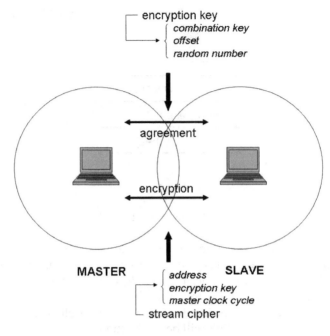

Fig. 5.7. Encryption.

5.7 Risk Potentials

The security risks in connection with Bluetooth traffic, the terminal devices and applications in question is partially similar to those known for WLANs and mobile phone usage. In addition, there are specific sources of endangerment brought about by Bluetooth operations and its own security mechanisms themselves. The following lists the most important problem areas:

- *Man in the middle attacks*: someone sneaks in between two communicating devices. This is facilitated if no data encryption is used.
- *PINs*: the main challenge is to employ PINs that cannot be guessed easily.
- *Tracking*: once the devices are switched on and running "Inquiry" they can be detected easily because of the polling process.
- *DoS attack*: by intense transmission of unsolicited information to one and the same address a device can be blocked.
- *Spying of data on terminal devices*: after successful spying local data can be read, manipulated or deleted.
- *Modification of the configuration*: there is special hacking software available able to access configurations and change parameters.

- *Bugs in the Bluetooth software*: known bugs enable non authorized access by attackers.
- *Default settings*: devices are delivered with default settings. These should be changed immediately after acquisition.
- *Long and frequent connections using the same key*: continually sending connection cipher keys and long standing connections themselves facilitate the decryption of that key by determined attackers. The pattern of these connections constitutes such a potential for an attacker.
- *Weak spots concerning the encryption algorithm*: as any other algorithm Bluetooth also has certain weak spots providing versatile attackers with possibilities to enter into communication.
- *Risk of theft*: because these are mobile devices the risk of theft is higher.
- *No password protection at the device level*: in case of missing passwords stolen devices can be utilised by attackers directly.
- *Malware*: as for any other network Piconets are targets for viruses, Trojan horses and relatives.

In addition there are risks specific to Bluetooth resulting from security holes in the standard itself. Here are the most important examples:

- **BlueSnarf:** enables an attacker to read data from the address directory or the calendar without the owner of the device knowing it. With suitable tools this attack can be carried out even in "invisible" mode. Usually a BlueSnarf attack is restricted to a distance of around 10 m from the victim.
- **BTChaos:** reads data with AT commands from a mobile phone. For this attack one needs a special C program and other freeware.
- **BlueBug:** functions similar to BlueSnarf. In this case AT commands can be transmitted to the victim device. These AT commands for example can initiate SMS. In addition—by using the SMS facility—the telephone number of the victim device can be obtained by BlueBug.
- **Bluejacking:** is employed in highly frequented places, such as railway stations, airports or trade fairs. Bluejacker defines the identification of a device with a special name, which is then displayed on the victim device via a connection request.
- **Backdoor:** in this attack the invader is able to manually switch his paired assault device to "invisible". Thereafter, the attacker is able to establish a Bluetooth connection without pairing request and without being detected by the victim.
- **BlueSmack:** a single Bluetooth PING package is capable to crash mobile phones. BlueSmack is a DoS attack similar to those known available on the Internet.

- **Bluetooth-Scanner:** the tool btxml.c creates a backup of personal data on a mobile phone. For some mobile phones this read out functions to the full extent. For these phones the telephone directory, the version number, the IMEI and all SMS can be read.

5.8 Countermeasures

The following countermeasures improve the security of Bluetooth applications:

- Procurement criteria

  There exist certain criteria, which need to be respected when purchasing Bluetooth devices. They comprise:
  - Minimal length of cipher keys
  - Possibilities to change default settings
  - Additional security software on offer by the manufacturer.

- Default settings
  All delivered default settings should be changed prior to the first deployment of any device.

- Services
  All service delivered but not to be used should be deactivated or uninstalled.

- Transmission power
  To reduce the risks of tracking the transmission power should be kept to a minimum.

- Security mode
  Security mode should be set to 2 or 3. Security mode 1 is out of the question.

- Encryption
  All communications should generally be encrypted. All data relevant to connections should be carefully stored in encrypted fashion on the devices themselves. There should be a separate policy for data encryption.

- PIN
  PINs should be composed out of all available character combinations (not only letters or numbers but also special characters in upper and lower case variations). The maximum length offered by the manufacturer should be fully used.

- Tracking
  It is difficult to neutralize tracking altogether as long as the devices are active. By combining several measures, however, security can be enhanced:
  - Switching the device to "hidden"
  - Changing the device designation
  - Deactivating devices not currently in use.

- Firewalls
  To protect against hacking software Bluetooth devices should be equipped with firewalls and other fences against viruses if technically feasible.

- Theft/loss
  In case of theft all cipher keys relevant for linking should be deleted on all remaining devices.

- Authentication
  Authentication procedures at the device level should be installed if technically feasible.

- Access
  If possible devices should be physically protected against unauthorized access.

5.9 Checklist

The following Table 5.5 regroups all relevant security aspects with regard to Bluetooth applications:

Table 5.5. Checklist Bluetooth.

Do you already deploy Bluetooth?	All Bluetooth applications in operation should be subject to regular security checks.
Is Bluetooth used in an organisation?	Bluetooth applications in organisations should be subject to internal security strategies.
Are you using Bluetooth as a private person?	Even in private environments risks are basically comparable technically to those in larger organisations.
Are you planning to introduce a Bluetooth application?	When implementing Bluetooth for the first time some general security aspects have to be taken into account.

Table 5.5 contd....

...Table 5.5 contd.

Are the responsibilities for security clearly regulated (strategically. organisationally, technically)?	The organisational security procedures comprise general organisational measures, technical measures concerning IT security, technical measures regarding communications.
	Unresolved responsibilities endanger regular operations.
Do the security guidelines contain a catalogue of countermeasures in case of security incidents?	Depending on the type of incident different countermeasures take effect.
	A classification of countermeasures should be obligatory.
Is the usage of Bluetooth subject to regulation?	Bluetooth applications should be taken into account by the internal security strategy.
	Unregulated usage left to the discretion of end users constitutes a serious risk potential.
Have security guidelines been documented regarding the usage of Bluetooth?	Bluetooth usage should be part of the organisational security concept.
	Users should be committed to the adherence to security standards.
Does an authorisation concept exist for applications?	An authorisation concept should be in place as part of user security as a matter of course.
	Without controlled access rights intruders will meet open doors.
Are applications protected by passwords?	Individual applications can be protected by passwords.
	Without passwords all applications can be accessed by anyone.
Is a password change strategy in place?	The change strategy comprises the rate of change as well as the password format.
	Change strategy should not be left to the discretion of the users.
Has the password change cycle been defined?	For the change cycle a time schedule should be communicated. Change should take place either monthly or at least quarterly.
	Passwords should be supplied with an expiry date.
Are complex password structures required?	The security of passphrases depends strongly on their length and character combination.
	Simple passwords can be guessed by an attacker easily.
Do alert procedures exist in case of security incidents?	Any organisation should have a suitable alert process in place.
	Without any agreed upon alert process reactions will result in activism.

Table 5.5 contd....

...Table 5.5 contd.

Are terminal devices inventoried?	The IT security officer together with IT management is responsible for documenting the existing infrastructure.
Is the infrastructure checked regularly?	This comprises physical inspection as well as checks on all relevant communication protocols and log files.
	Checking the records should follow a regular schedule.
Are configuration responsibilities sorted out?	Bluetooth adjustments can be done from basically any workstation.
	Restrictive measures have to be taken to allow configuration only be administrators.
Is security mode 1 in place?	Mode 1 does not use any special security mechanisms and no authentication procedures.
	Mode 1 should in no case be in operation.
Is security mode 2 in place?	This mode provides security mechanisms at the service level.
	Mode 2 should be the minimum standard.
Is security mode 3 in place?	This mode provides security procedures at the link level (encryption of messages and data).
Are cryptographical procedures used to authenticate access to the Piconet?	The basis of those cryptographical procedures are connection cipher keys in combination with a pairing mechanism between devices.
	Lacking authentication by Piconet subscribers will facilitate access by non-authorized third parties.
Are data encrypted?	Encryption can proceed only after authentication. For this special cipher keys have to be agreed upon.
	Not encrypted data can be easily spied upon.
Are data on local storage media encrypted?	Bluetooth security provisions end after the communication process.
	For encryption on local media the usual organisational standards should take effect.
Are complex PINs in use?	This constitutes a main problem, once PINs are used that can be guessed easily.
	Maximum PIN length should be used together with special characters.
Are devices switched on, even when they are not in use?	Devices should be switched off, when they are not in use.
	Once devices are switched on they can be localised easily because of the polling mechanism.

Table 5.5 contd....

...Table 5.5 contd.

Are the standard security settings of the devices kept in place?	Standard settings should be changed immediately after purchase.
	Information about standard settings are publicly available.
Are frequent connections or connections of long duration using the same cipher key common practise?	Continual usage of the same cipher key facilitates the discovery of this key by determined attackers because of these connection patterns.
Are devices protected by passwords?	Devices can also be protected by passwords.
	Without password protection stolen devices can be used directly by attackers.
Have procurement criteria been defined for Bluetooth devices?	They should take into account the security relevant features of manufacturers.
	Those include: minimal length of cipher keys, change provisions of pre-adjustments and other special security software.
Are services not in use deactivated?	Services included in the delivery but not intended for use should be deactivated.
	Any service constitutes its own risk potential.
Is transmission power kept to a minimum?	There exists a compromise between desired and security relevant range.
	Because Bluetooth devices are continuously polling transmission power should be kept to the utmost minimum necessary.
Are provisions against tracking in place?	By combining certain measures security can be enhanced.
	These include: change of device designation and deactivation when out of use.
Is the device number changed?	The device number can be changed with the help of the configuration software.
	By changing the device number tracking will be made more difficult.
Are Bluetooth devices protected by firewalls?	On Bluetooth devices firewalls can be installed.
	Firewalls are standard features of any security philosophy.
Does a master plan exist in case of theft?	After theft all connection cipher keys have to be deleted on the remaining devices.
	Mobile Bluetooth devices are especially prone to theft or loss.
Are devices physically protected against external access?	If possible devices should be protected by physical means against unauthorized access.
	Mobile Bluetooth devices are especially prone to theft or loss.

6

Infra Red

〈〉

6.1 Background

Besides wireless applications based on WLAN and Bluetooth technologies communication via infrared radiation is on offer since some time. Infrared is light with wavelengths between 7.8×10^{-7} and 10^{-3} m corresponding to a frequency range of 3×10^{11} Hz up to 4×10^{14} Hz. One advantage of infrared radiation is its marginal harmfulness and susceptibility against electrical interference. One disadvantage is its small reach. Other advantages include:

- Simple and low cost implementation
- Low electrical power requirements
- Directed point to point connection
- Efficient and reliable data transmission.

For infrared communication standards have been developed which are oriented to possible applications. In the following these standards will be presented with regard to their architecture including transmission protocols. Thereafter, possible applications will be considered and potential risks discussed. A small checklist at the end of this section lists key points to be observed when choosing and implementing infrared communication.

6.2 IrDA

Infrared has already been in use as transmission medium for some time for controllers, printers, pocket calculators, and smartphones. In 1993 HP, IBM and Sharp initiated a group called Infrared Data Association (IrDA) [98] to promote the development of an industrial standard for infrared communication. Already in 1995 the first products adhering to this standard were launched on the market. They comprised:

- Notebooks equipped with an infrared interface
- Printers
- Infrared adapters for PCs.

Contrary to its predecessors, which use proprietary protocols, devices adhering to IrDA are capable to communicate between different applications on hardware from diverse manufacturers.

Table 6.1 contains different data rates corresponding to a reach of about 1 m in line of sight corresponding to the relevant IrDA protocol [99] specifications:

Table 6.1. IrDA Data Specifications.

IrDA Data Specification	Transmission Rates [KBit/s]
SIR	9,6–115,2
MIR	576–1152
FIR	4000
VFIR	16,000
UFIR	96,000

6.2.1 General Considerations

Figure 6.1 shows a systematic comparison between classical wires bound against infrared connection [100]:

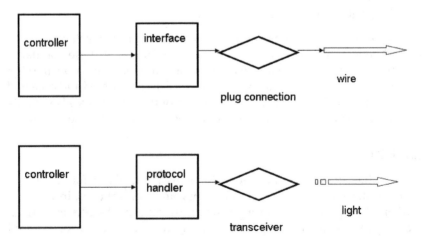

Fig. 6.1. Wire Against Infrared.

There are two decisive elements:

- The protocol handler and
- The optical transceiver (sending and receiving unit).

Figure 6.2 shows the schematic interplay between a peripheral device —in this case a laptop—and the interface of another system:

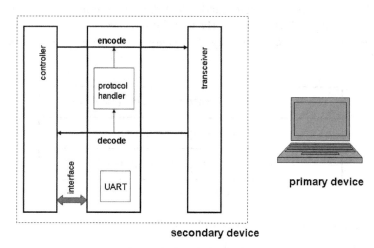

Fig. 6.2. Infrared Connection.

From right to left the following items are visible:

- Laptop
- Optical transceiver
- Protocol driver
- Host controller
- Universal asynchronous sending and receiving unit.

The most important layers of the protocol according to the OSI standard are:

- Physical Layer
- Link Access Layer
- Link Management Layer
- Application Layer.

Link Management and Application Layer themselves have again substructures. In the following these protocols will be discussed in more detail. All layers are implemented in the protocol driver.

6.2.2 Protocol

The Physical Layer resides at the bottom (Fig. 6.3):

asynchron Serial IR (SIR) 9600-115200 baud	synchron Serial IR (SIR) 1,15 Mbaud	synchron Fast IR (FIR) 4 Mbaud

Fig. 6.3. Physical Layer.

The Physical Layer fixes the data format. Up to three specifications from Table 6.1 can be implemented. Most devices employ SIR (Serial IR). PCs and some printers need FIR (Fast Serial IR).

The next layer is the Link Layer. This layer determines the connection type (Fig. 6.4):

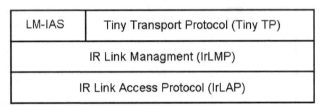

LM-IAS	Tiny Transport Protocol (Tiny TP)
IR Link Managment (IrLMP)	
IR Link Access Protocol (IrLAP)	

Fig. 6.4. Link Layer.

The Link Layer is divided into sub layers:

- Link Access Protocol (IrLAP)
- Link Management (IrLM)
- Optional transport protocols.

The layers provide for:

- Data Routing
- Error Corrections in Data Packages
- Link Management
- Information Structuring for the Application Layer

of the protocol stacks. On top of all resides the Application Layer (Fig. 6.5).

This is, where all the different application protocols reside. Here the factual object transmission (files, programs, photos, etc.) is managed. The characteristics of these objects have to be defined beforehand.

IrTRAN-P	IrObex	IrLAN	IrCOMM	IrMC

Fig. 6.5. Application Layer.

IrCOMM stands for IrDA Standard Specification, which replaces the traditional serial and parallel interfaces.

6.3 Applications

If IR communication shall be implemented for certain terminal devices technical requirements have to be checked and—if necessary—certain features to be installed. Implementation itself is relatively straight forward. Communication proceeds according to protocol (s. below).

6.3.1 Terminal Devices

IrDA interfaces are available on:

- Notebooks
- Mobile phones
- Printers
- Pagers
- Special watches, to measure heart beat for example.

6.3.2 Preconditions

To facilitate IrDA communication for notebooks, PCs or smartphone, a digital interface is required as well as an analogue front-end component. The latter can be connected via an RS-232 serial port for up to certain transmission rates (SIR) or via a USB adapter. Many devices on the market already have built-in infrared ports: laptops, mobile phones. In addition the necessary driver software has to be found on the device.

6.3.3 Communicating

Figure 6.6 shows the connecting sequence according to the standard IrDA protocol:

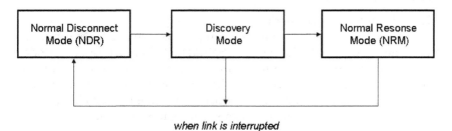

when link is interrupted

Fig. 6.6. Connection Sequence.

Three modes are possible:

- Normal disconnect mode
- Discovery mode and
- Normal response mode.

These modes shall be discussed now:

6.3.3.1 Normal Disconnect Mode

Figure 6.7 illustrates this mode:

primary device

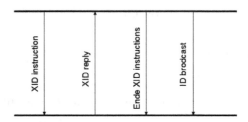

secondery device

Fig. 6.7. Normal Disconnect Mode.

NDM is the mode under which a device is polling for other IrDA standard devices. In this case the device sends XID (exchange identification) instructions within a time window of between 0 and 7 seconds. As soon as a second device is within reach of the first it will reply and reserve the time window. After this the second device will ignore all following XID messages. The IrDA protocol allows distinguishing between up to eight different other devices. The first device emits a broadcast ID to which the other device does not respond.

6.3.3.2 Discovery Mode

Figure 6.8 illustrates the mode:

Under this mode the communicating devices negotiate their mutual parameters. The first device sends an SNRM (Set Normal Response Mode) instruction together with certain parameters and connection addresses. The other device sends a UA reply with parameters using the mentioned address. Thereafter the first device opens up a channel for IAS queries, which the second device then confirms. The features include:

- IR baud rate
- Size of data packages
- Delivery time and others.

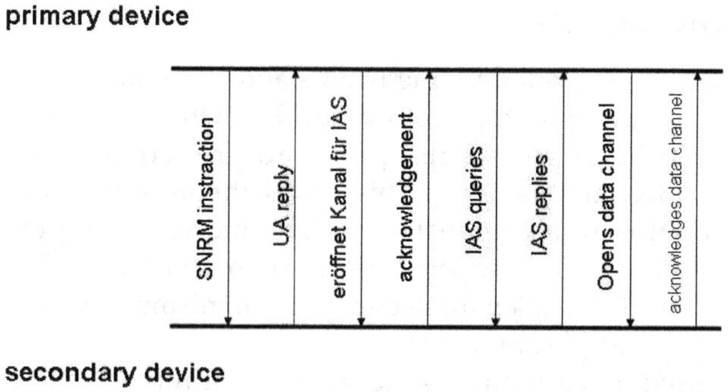

primary device

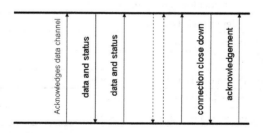

secondary device

Fig. 6.8. Discovery Mode.

These parameters are now being exchanged to find the highest common denominator to optimise performance. Now the first device has to come forward with the request for data. For a smartphone this happens once the first data package for transmission is ready. For PCs a special program has to be installed to manage the IR port. The second device confirms that the channel is now open for data.

6.3.3.3 Normal Response Mode

Figure 6.9 illustrates this mode:

NRM is the mode under which data and control information are sent to and fro. Status information is important to find out, whether a connection is still active and not blocked. Should the connection be blocked because a time out has been reached the device is reset to the NDM state. As soon as communication is finished the first device disrupts the connection. The second device confirms this and both return to the NDM state.

primary device

secondary device

Fig. 6.9. Normal Response Mode.

6.4 Security Aspects

Security checking provided by the IrDA standard is only concerned with technical safeguarding at the protocol level—such as transmission errors [101]. Authentication procedures, password protection, and encryption are not provided at this level. This could theoretically mean that IR communication is weak against recording and eavesdropping. In this respect, these are higher security risks as compared to classical LAN applications. If at all, relevant security mechanisms are required to be introduced at the application level.

One consequence would be to avoid continuous operation of IR interfaces, since otherwise non-authorized persons could send data to a device using such an interface. For different devices there exist different risks:

- Laptop: data and programs
- Mobile phones: SMS data and programs

All these pieces of information could theoretically be infected by malware.

An additional aspect concerning security can be regarded as positive and is due to the fact that IR communication has a short range and thus can only happen within a reduced area. There is, however, a residual risk emanating from radiation scattering of the communication components.

6.5 Checklist

Table 6.2 contains a short checklist regarding the deployment of infrared.

Table 6.2. Checklist IrDA.

Check Item	Reply
Do you consider the deployment of infrared?	
Which types of devices will be chosen?	
Which applications are planned?	
Who are the users?	
Which transmission rate is under consideration?	
Will the devices be linked to other networks?	
Are the locations suited for infrared radiation?	
Are the locations protected against the detection of radiation scattering?	
Are the devices equipped with the necessary drivers?	
Are the devices equipped with the necessary infrared interfaces?	

7

Near Field Communication [102]

◇◇

7.1 Introduction

Near field capable devices, such as tablets and smartphones are sold everywhere today but seldom used for this purpose. The feature "Tap and Send" is the one mostly employed. For this to function one has to approach two devices in a way that their NFC-antennas touch each other or stay separated for just a few centimetres. The receiving device offers according to different apps selected information, triggers an app or executes a data transfer. This can be controlled by "accept" or "reject" functions.

Besides this, read and write of NFC tags is the second field of usage of this technology. Such tags are on offer in many variations, such as adhesive labels, key fobs as plastic cards, etc.

This chapter will cover the technology and possible applications for smartphones and tablets.

7.2 Technology

7.2.1 History

The Near Field Technology is based on the RFID (Radio Frequency Identification) used in goods security systems in the seventies of the last century. RFID very quickly spread to access control systems, cashless payments, and further on to immobilisers. The hardware consists of a reading device and a passive transponder. Passive means in this case that the transponder does not require its own power source. In 2004 Phillips, Sony, and Nokia founded the NFC Forum [103], which developed the necessary specifications. In 2006, Nokia released the first mobile phone with NFC technology on the market, and in 2011 this technology became part of the operating system Symbian [104]. Samsung started in 2011 to distribute an NFC capable smartphone under Android. Only during 2012 did Microsoft provide smartphones with NFC under Windows Phone 8.0 and Windows 8.0 RT edition. Since RT functions are available for programmers under Windows 8.0 Desktop, even desktops programs can employ NFC once the device contains an NFC chip. This is at the moment (2018) only the case for Windows tablets, some Ultrabooks and convertibles.

7.3 Specifications

NFC is a wireless technology in the 13.56 MHz range requiring little power. The transmission speeds are 106, 2121 and 424 Kbit/s, which is not very fast. To communicate an initiator, an active element and a target object as passive element are required. Two active elements together can build up a peer-to-peer (P2P) connection. For example two smartphones can exchange data by a "Tap and Send" function; in this case the receiving device cuts off its current until the end of reception.

Theoretically, the distance between sending and receiving devices can be up to 20 cm, practically, however, this ends at about 4 cm. But this feaure makes the transmission very secure against wire tapping. Table 7.1 shows the differences between NFC, Bluetooth normal and Bluetooth Low Energy and the corresponding technical specifications.

Table 7.1. NFC and Bluetooth.

Feature	NFC	Bluetooth 3.0	Bluetooth Low Energy
Passive RFID Compatible[1]	ISO 18000-3	activ[2]	activ
Network	ISO 13157	IEEE 802.15.1	IEEE 802.15.1
Network type	point to point	WPAN[3]	WPAN
Range	< 20 cm	ca. 100 m	ca. 50 m
Bit rate	424 Kbit/s	2,1 Mbit/s	25 Mbit/s
Frequency	13,56 MHz	2,4–2,5 GHz	2,4–2,5 GHz
Interlinking time	< 0,1 s	< 6 s	< 0,006 s
Power consumption	< 15 mA reading	depends on performance class	< 15 mA reading + writing

Legend:
[1]Passive RFID: Radio Frequency Identification with power-off devices
[2]Active: current-on
[3]WPAN: Wireless Personal Area Network

7.3.1 Active Applications

7.3.1.1 Example: Transmission of a Photo

Active applications are such that both NFC devices have power-on, for example two smartphones or a tablet and a smartphone. Both devices have to be activated, that is, the blocking screen has to be switched off resp., a user has to be logged in. This feature has to be activated in both devices.

The following sequence of steps has to be executed (for example under Windows 8.1):

1. Selection of photo from the store of the sending device.
2. Selection of function "Send" from pop-up menu.
3. From next menu select "Tap and Send".
4. After the prompt "Tap" approach the sending device to the receiving device.
5. The receiving device is then asked whether it wants to receive the proposed content.
6. Thereafter, the transmission proceeds and the image is stored on the receiving device (the destination index notation depends on the version of the operating system).

7.3.2 Technologies

Even between active devices NFC can transmit only a few kilobytes data. These data have to conform to the NFC format with its different record types. This is why photos, videos, office documents, etc., are transmitted via Bluetooth, NFC takes over the pairing of the devices involved and suspends these after the end of the transmission.

Unfortunately these features are limited to Windows only. Apple's iOS does not yet own an NFC. The Android of Google has been developed even further than Windows, and identifies the device, but is not capable for pairing. The reason may be the different URIs (Uniform Resource Identifier) used for start-up.

In the meantime ear phones, speakers, hands-free kits, etc., for NFC and Bluetooth capable devices are on offer, which have to be touched only slightly for pairing to initiate a connection. These devices have batteries allowing up to 15 hours of operation.

A further possibility for pairing is via Wi-Fi or WLAN. Samsung has developed it for Android devices via S-Beam. Windows 8.1 offers this possibility only for printers equipped with NFC but only one-directionally. Windows 8.1 and also Windows phone 8.1 will serve as platforms for these applications using NFC pairing to exchange large documents, videos, photos, etc., via Wi-Fi.

7.3.3 Passive Applications

This section deals with devices without own power supplies, fed by active NFC via antennas. Device means in this connection so called tags like adhesive labels, bracelets, tags, chips, and business cards or smart poster. There are even mobile phone cushions with re-writable tags on each side on offer.

7.3.4 NFC Forum Specification

The NFC Forum was founded in 2004 and has since released international guidelines for four type specifications for different tags, which define characteristics, such as memory sizes for example. They are based on existing contact free commercially available products and allow reading, writing and formatting resp. the prohibiting of writing and formatting, once devices conform to these guidelines. Furthermore, they define minimum and maximum memory sizes.

Furthermore, the forum has also fixed the format and the rules for Record Type Definitions (RTD). These definitions are based in turn on the NDEF (NFC Data Exchange Format), with which nearly all tags are formatted. These record types consist of five basic types and can be expanded by users. The basic types are:

1. NFC-Text for texts in many languages
2. NFC-URI for Uniform Resource Identifier. With this tags can be started via programs, settings changed and web pages called.
3. NFC_Smart Poster for saving web pages, SMS texts or telephone numbers.
4. NFC-Generic_Control has been withdrawn without substitution.
5. NFC-Signature served for signing individual or selected records. It can be specified, whether a signature is obligatory or optional.

Table 7.2. NFC Guideliness.

Type	Guideline	Reading + Writing	Only Reading	Capacity
1	ISO/IEC 14443A	yes	configurable	96–2000 Bytes
2	ISO/IEC 14443A	yes	configurable	48–2000 Bytes
3	(JIS)[1] X 6319-4	yes	pre-confugured[2]	limit 1 MByte
4	ISO/IEC 14443	yes	pre-configured	limit 32 KByte

Legend
[1] = Japanese Industrial Standard also called FeliCa [105]
[2] = pre-configured by the supplier and not alterable

In the meantime from these types the following standards (selection) for mobile devices have been created:

vCard (business card): platform support for WP8, Android, MeeGo, Symbian and BlackBerry. Recommended are 1 Kbyte at a minimum.

URI: platform support for all mobile NFC devices. Example: https://, http://www or "tel:". WP8, Android and Windows 8.1 also support the launch of applications and device settings. The launch of applications on alternative platforms is at present not possible.

Geo: tag link format for latitudes and longitudes in decimal representation according to WSG-84 (World Geodetic System 1984). In WP8 these positions can be linked to a map program in a way that the map shows it as centre point or takes this position from the map and writes it to the record.

Smart Poster contains for example the URL of a website and head lines or texts in several languages and pre-ceded by a country code. Users then obtain leads to the web page in their language.

Email calls the email editor and enters a stored address and other information.

Text stores normal text. But Windows and Android do not link any application with this text.

7.3.5 Example Applications

When buying NFC tags it is important to check that these are formatted in NDEF, since Windows—contrary to Android—offers no possibility to format tags.

7.3.5.1 NFC Interactor

This application had been developed by using the NFC/NDEF open source library in C# and offers a multitude of possibilities to analyse, read, write to, and clone NFC tags. The following example shows how to write to a pre-formatted tag offering to a user to search for an application in the Windows phone store and, if found, to also install it (Fig. 7.1).

In the tag composer choose "tag type" to write. It will open the selection window as can be seen in Fig. 7.2. By activating the interrogation mark, the individual records will be explained.

After selecting the indicated element on the left, one has to enter "DB" on the right and press the button "Search". Thereafter, all Apps starting with "DB" will be shown. "DB Navigator" is selected (Fig. 7.3). By pressing the button on the left hand side at the bottom, the image on the right appears asking to approach an NFC tag to the phone. Once the tag OS is recognised as writable, an acknowledgement is sent. If required, further tags could be

Fig. 7.1. Interactor Start Screen and Tag Composer.

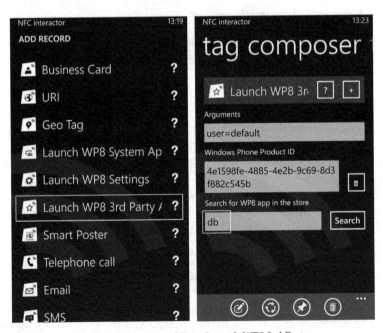

Fig. 7.2. Selection and Type Launch WP8 3rd Party.

successively overwritten with the record. To leave the process and return to the menu one has to press the yellow arrow key.

Before leaving the program one has to wipe to the right to obtain the tag just written to on the phone.

Fig. 7.3. Search Results.

7.4 Security Aspects

Although NFC recognition takes place in a circumference of only 5 cm, this area can be enlarged for example in stores to about 1 m. This leads to the following potential risks:

- NFC records cannot be encrypted. Authentification against a tag is not possible.
- Overwriting of records by a jamming transmitter or content reading with special devices
- Loss of the device and or tag.

Since Windows Proximity APIs do neither authenticate, encrypt nor guarantee the integrity of tag contents, the exchange of sensitive data via NFC should be avoided. These include:

- Passwords
- Private photos

- Financial data
- Emails
- Social security numbers

   to cite but a few.

Some applications provide a password to the tag but encrypted and only usable with the device in question. This forces the user to take special care of the device and not let it pass into unauthorized hands.

7.5 Conclusions

NFC is gaining in importance in modern computer technology, especially for tablets, smartphones and devices offering NFC support like loud speakers and printers. Unfortunately, no binding standard has been enforced yet but only recommendations, to which most suppliers adhere. In several countries projects do already exist or are in test for small amounts cash payments or ticket purchase. MasterCard, Visa, and other credit cards are working on solutions for payments with cards containing NFC chips.

In April 2014, the Canadian enterprise Rogers has released SIM cards with NFC chips for the regularisation of amounts between 50 and 100 dollars. After a user enters a pass code he can approach his smartphone to a payments terminal and the purchase is closed. In Canada, already many supermarkets and businesses have been equipped with such terminals. Today, there are prepaid cards from MasterCard and various gift cards, which customers can download onto their mobile phones. Because of safety considerations in Europe these services may still take some time to arrive.

Security Policy

In the preceding chapters a typical directive has been introduced (for mobile phones). The comprehensive directive proposed in this chapter contains some general elements, which can serve as the basis for any particular directive in an organisation. For specific applications templates are presented, which have to be completed in the appropriate places. To do this, contents from the preceding chapters in this book could be used.

8.1 Introduction

Wireless Security is part of a comprehensive concept regarding IT security. The latter can be divided into numerous documents all being mutually referred to each other and their complexity depending on the installation in question. It can be distinguished between strategic, technical, and organisational measures and their corresponding directives. According to these levels the group of people concerned with these instructions varies as well. A directive for the design of firewalls is of no interest to the common user. He wants to know what the structure of his passwords should look like.

8.1.1 Security Requirements

Security requirements are visible on different levels of relevance:

- At the strategic level and its relation to the overall organisation
- As tools to satisfy certain specifications
- Through groups of people, who are responsible for specifications.

These different dimensions will be taken care of in the following. They are again related to specific risk estimates and possible countermeasures.

8.1.2 Risks

Risks again can be categorized multi-dimensionally:

- According to objects
- According to potential damage
- Or as combination of both.

Additionally, risks vary depending on the progress of attacks: the further an invader progresses in a system the higher will be the remaining risk. Risks can never be completely eliminated. Aim and subject of this directive is to minimize all possible risks one can think of. For those risk potentials, which will follow later, and which are technology dependent, plausible risk scenarios will be developed together with relevant pre-emptive and compensational measures.

8.1.3 Measures

As explained further down one has to distinguish between two categories of measures:

- Organisational and
- Technical.

Both operate in concert and complement each other. Measures can be of general nature, which constitute a security environment and which generally controls security loopholes. These include directives, organisational structures and technical security installations at hardware and software level. Additionally, quite a number of specific measures exist, which cover specific security risks and are relevant for specific cases. The relevant processes have to be implemented. Such measures will be discussed in detail.

8.2 Scope

The scope of IT security is limited by two criteria:

- Organisation
- Time.

The scope concerning organisation refers to the organisational units in a company, for which this system and its documentation are relevant. Normally all units are included. Exceptions may be outsourced units, subsidiaries or affiliated companies. In times of transition after fusion with other companies for example the possibility exists that certain departments, which may be using different IT systems, are controlled in a different fashion. These exceptions have to be documented properly.

Timely restriction of the scope refers to version levels. Every document has a version number referring to the main document. The validity statement refers to the actual version, exceptionally also to sections of past versions. In any case: the latest update is valid. This comprises statements as to how individual documents are processed. Changes are to be recorded in a version history up to the final release.

8.2.1 Normative References

The whole complex of subjects concerning IT security is again influenced by national standards and directives, some of which will be briefly mentioned. Detailed information can be obtained from the original documents.

8.2.1.1 Legal Regulations

Every country issues laws, which may also be pertinent to IT security under different aspects. These include:

- Data privacy protection
- Laws regulating information and communication services
- Laws regulating telecommunications
- Signature regulations
- And many others.

8.2.1.2 Guidelines and Standards

Government agencies offer guidelines based on international standards concerning IT security. Three of these standards will be outline in the following:

8.2.1.3 Standard ISO/IEC 13335 [106]

This standard together with the two others presented here were developed in cooperation with the International Electrotechnical Commission in Geneva. This document outlines general principles as a reference base for more specific standards. It mainly contains:

- Concepts and models for security in information and communication technology
- Technical preconditions for the management of security risks
- Guidelines concerning network security.

8.2.1.4 Standard ISO/IEC 17799 [107]

This standard offers approaches and step sequences for the strategic implementation of IT security systems. Detailed technical instructions are not included in this document. Its character is recommendatory without any binding force.

8.2.1.5 Standard 27001 [108]

The title of this standard reads: "Information Technology—Security Techniques—Information Security Management Systems Requirements Specifications". This standard also has recommendatory character. Technical instructions for implementation are not given.

8.3 Information and Communication Security

IT security plays a major role, when systems are implemented. It can be regarded as a stand-alone subject or be part of IT quality management in general. Even if a separate IT quality organisation exists, mutual interconnections and dependencies are so multiple that one cannot be considered without the other. IT quality management is a precondition for a clean implementation of security aspects. On the other hand, without taking security aspects into account there will be no sustainability in quality control.

Figure 8.1 shows the relationship between IT quality management and IT security in an overall project:

The compliance of security requirements depends on the strategic placement of these tasks in a company as a whole. It will be outlined, which preconditions have to be initiated for this role. The involvement of all project members at all times is important.

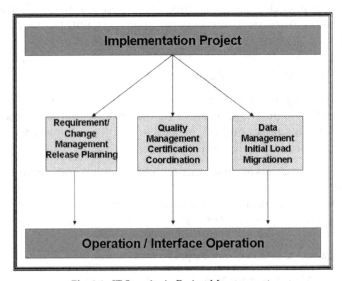

Fig. 8.1. IT Security in Project Management.

8.3.1 Strategic Involvement

IT security management is part of the overall security management covering the whole company—comprising all other material and immaterial goods as well as all employees. As such IT security management should be integrated methodically and process related into the overall concerns of security management. When talking about IT security management it is understood that the whole complex of IT and communication security is addressed.

IT security is achieved by a number of conceptual and organisational measures as well as the necessary technical preconditions, which are relevant to reach defined security objectives. These are the areas of concern:

- IT processes
- Computer systems
- Hardware
- Software
- Communication installations
- Data
- Documentation.

Security and Safety Engineering is the platform, on which technical preconditions for IT security can be created. The requirements can be derived from security criteria specific to a company, fixed by management after consulting with security experts. Among them are such classical criteria as:

- Data integrity
- Confidentiality, etc.

together with for example availability and authenticity. If a company deploys wireless communication networks, these criteria will look different from those for pure LAN applications. The principle has to be mutually agreed at the top level and communicated as binding security policy. Security policy should be positioned as part of the company's guiding principles, and should be furnished with the necessary competencies at top management level.

On the basis of these definitions documents structured in hierarchical manner are drafted on the various execution levels, transforming these guiding principles into directives to be filled with life.

8.3.2 Security Organisation

As a matter of course all employees and therefore all members of a project team have to be briefed about all valid security directives in a company. This may happen at the instant of providing an email account for example, and by then sending the relevant information. In special cases, such as working with a WLAN the persons concerned should receive the necessary instructions separately. Instructing administrators should be obligatory in any case, since these persons have access to sensitive company data and configurations. Security aspects concerning administrators normally exceed those of common users.

After successful training, instruction and receiving the relevant security documentation every employee has to acknowledge by his signature on a special form that he has been informed, that he agrees with the directive and will respect it. The signed acknowledgement has to be archived by the IT security organisation.

Table 8.1 recapitulates the strategic preconditions to constitute an IT security management.

Table 8.1. Checklist IT Security.

Does an IT security management exist?	IT security management deals with all security aspects concerning implementation and operation of IT installations.
Have the concerns of IT security management been documented?	Precondition for an effective IT security management is the relevant documentation.
Are current IT standards taken into account with regard to security management?	ISO/IEC 13335, 17799, 27001
Have IT security criteria been documented?	Security is classified according to such criteria as confidentiality, availability, integrity, etc.
Will the participants of IT security trainings acknowledge their participation by signature?	The participation in security trainings should be documented in the interest of all concerned.
Is the adherence to security directives monitored regularly?	The monitoring of the adherence to security directives should follow an appropriate action plan.

8.3.3 Approval Process

Organisational procedures have to be introduced to secure the approval of different services or objects, including:

- Allocation of accounts
- Access authorisation to applications
- Control over terminal devices.

Normally three instances are concerned with this process:

- Applicant
- Supervisor
- Clearing officer.

The transaction has to be documented and downstream organisational units have to be informed (controlling, procurement, etc.). Once the applicant leaves the organisation all authorizations become invalid and have to be withdrawn.

8.3.4 Confidentiality

Another important transaction to improve the security of an organisation is the commitment to confidentiality. Generally, such a commitment is governed by the work contract such that no separate documents have to be drafted. Additionally these regulations are still valid for the time after a person has left an organisation. However, occasionally the need arises for specific confidentiality instructions. This can be the case for example, when a person gets into contact with highly sensible data while working for a specific project. In such cases the confidentiality commitment may include restrictive information policies against units and persons internal to the organisation. Sometimes the signature under an appropriate paper may be necessary. And this may not only concern data. Sometimes reports about internal processes facilitate inferences about methods of payment, applications, etc.

8.4 Physical Security

Besides the security problems directly connected to information and communication technology itself discussed further on there are the normal security aspects concerning buildings and equipment, which in most cases have to be solved physically.

8.4.1 Objects

These are some of security relevant objects:

- The whole area of a company or other organisations
- All buildings; and especially rooms that have direct communicative access to computer systems and communication installations
- Utility services
- All hardware in conjunction with information and communication, mobile or fixed
- The adjacent neighbourhood of the company grounds, in as much as access to internal systems may be attempted wirelessly.

All these installations have to be secured in different ways, once the potential for direct impact exists.

8.4.2 Access

The first and most important obstacle against non authorized access is the selective accordance of admission to the installations of an organisation. This subject will not be covered in detail here, since admission control is a science in its own right. It is important that always all currently available technologies be used to secure all rooms, which house central hardware for application systems by special admission mechanisms within or in addition to the already practised admission security to the premises themselves.

Terminal devices, which are placed in offices, should be physically fixed and switched off, when offices are deserted.

8.4.3 Threats

As will be outlined further down possible threats are manifold and specific for the area of wireless communications and surpass classical risk scenarios. They can be classified roughly in the following manner:

- Direct access to central hardware with the intention to destroy or disrupt operations
- Spying attempts on central applications
- Spying attempts on decentralised applications
- Attempts to manipulate central and/or decentralised data
- Deployment of malware
- Theft of terminal devices.

In addition there are other aspects to be followed up.

8.4.4 Equipment

Equipment commonly being subject to high risk potential are among others:

- Central IT installations
- Fixed peripheral devices
- Mobile terminal devices
- External storage media
- Communication modules (modems, ports, switches, etc.)

8.4.5 Utility Services

Utility services are prone to create risks, once they:

- Do not function or
- Function wrongly.

Electricity supply belongs to the first category. To prevent interruptions emergency power supplies have to be on standby. Water supplies belong to the latter category, if large quantities of water penetrate computer rooms and endanger hardware due to water pipe fracture. For both cases emergency plans have to be drawn up.

8.4.6 Disposal

Besides the usual legal disposal regulations special attention has to be drawn to additional aspects concerning company and IT security:

- Prior to disposal all data stored on devices—most important administration data—have to be deleted or neutralized in such a way that even accomplished technicians will not be able to re-constitute them.
- Indications to the company like type labels and inventory labels have to be removed. In this way inferences about the original company, where they were in use, should not be possible.

8.5 Documentation

Here are the most important elements to be considered for individual directives:

- Subject of the directive (hardware: laptop; software: intranet for example)
- Application procedures for usage
- Responsibilities for usage and costs
- Limitations of usage and costs
- Interdictions
- Liability
- Damages.

Directives are of a general nature or relevant to specific fields of technologies. One has to distinguish between the proper directive itself and the corresponding rules of implementation.

8.5.1 Processes

Quite similar to other aspects of IT quality management the Deming Process [109], so called after the famous American quality guru W. Edwards Deming, plays an important role for IT security philosophy with respect to verification, compliance and evolution. Figure 8.2 shows this process schematically:

Always the same cycle has to be passed:

system constitution > implementation > analysis > improvement

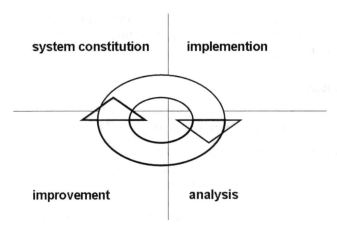

Fig. 8.2. The Deming Process.

The policy outlined so far has to be prepared organisationally and technically. Thereafter implementation of it takes place together with all parties concerned. After a certain time of operation experience is gained resulting finally into new proposals and improvements, and the whole process starts all over again. It is important to note that the run time phase is not equal to a trial phase. In fact this is a continuing process with fixed review intervals. And its aim is not only to correct initial design mistakes. The overall process should rather make sure that especially within the IT environment the newest technological developments are taken into account regarding security aspects.

8.5.2 Commitment

A policy or directive without sanctionable commitment only possesses its paper worth, on which it is printed. Legally there are a number of possibilities to ensure the compliance with such a document.

8.5.2.1 Non-compliance

Notice to the IT security officer.

8.5.2.2 Acknowledgement of Instructions/Sample

This directive should be part of comprehensive employee security instructions. At the end the following agreement can be signed:

Acknowledgement of the Directive

"Please read the present .... directive and countersign it at the bottom of the document. One copy with your signature will be kept by the IT security officer.

With your signature you acknowledge:

- I have received the .... directive, understood its meaning and agree with it
- Confirmation of individual requirements from the directive
- Confirmation of the confidentiality clause
- Confirmation of liability and responsibility
- I understand that non-compliance regarding this directive can induce legal consequences
- Name
- Signature
- Department
- Date

8.6 Wireless Security

Now elaborations from the preceding chapters can be brought in, for example:

- Open and protected networks
- WLAN generations
- Security requirements
  - Safeguarding availability
  - Securing data integrity
  - Securing authenticity
  - Securing confidentiality
  - Security risks

- Mobile Phones
  - General risk potentials and strategic countermeasures
  - General organisational countermeasures
  - General technical countermeasures
  - Distinctive risk potentials concerning mobile phone communication
  - Attacker holding a terminal device
  - Attacker not holding a terminal device
  - General protective measures

- Bluetooth
  - Instruments
  - Risk potentials
  - Countermeasures

- Infrared
  - Risk potentials
  - Countermeasures

8.7 Summary

This general directive is the basis for other instructions and directives, which have to be worked out on a case to case basis for certain groups of people using certain technologies. Individuals concerned should be asked to commit themselves to those directives with regard to their fields of activity. The commitment takes place after instruction and training by signature of this separate document as outlined above. This present directive documents the strategic thoughts with respect to security of an organisation, for which it has been drafted and as such is part of the overall IT or other security strategy and thus part of the company strategy as a whole.

9

Emergency Management in Communication Networks

◇◇

9.1 Emergency Management Systems

Emergency management systems are not only relevant once a real emergency arises, but they also serve to prevent them or to prepare for crises and emergency scenarios. In this way preparatory measure are defined to minimize the impact of sudden emergencies with respect to the core processes of an organisation (administrations, enterprises) and to provide for prompt resumption of normal activities. To arrange and plan these preparations for all eventualities in a suitable manner these processes and the relevant staff and systems have to be identified beforehand.

9.1.1 Why Emergency Management?

Besides the pure interest to continue with daily business and, therefore, existentially to maintain the business as such there are other concrete reasons to conceptualise emergency management. Although there are basically requirements by law for emergency management as such there are indeed legal and contractual obligations resulting from the purpose of an enterprise. These may comprise contractual obligations for the fulfilment of deliveries and services agreed between clients and an enterprise. In certain classes of business the necessity for such measures are deduced from other laws. This is the case for example for banks but also for corporations listed at the stock exchange being controlled by laws guaranteeing transparency. These corporations have to comply to risk management. Furthermore, there are other laws making emergency management a necessity (work safety, financing, etc.)

9.1.2 *What is Emergency Management?*

In a nutshell emergency management can be defined as follows:
Emergency Management is

- A systematic approach oriented on business processes
  - To limit exceptional circumstances
  - To limit the effect of damages arising from unforeseen external and internal impact.

- The installation of organisational requirements comprising:
  - An organisational structure already active in preparation of the definition of preventive measures but in part only activated in case of emergency
  - A process organisation to be activated in an emergency.

- The development of relevant concepts in line with the strategic objectives of an organisation and its core processes
- Quick reaction in emergencies by employing the measures previously defined
- Enabling the continuation of the most important business processes under the circumstances left by an emergency.

9.2 Standards

9.2.1 *ISO 22301[110]*

The ISO 22301 standard represents the most recent guideline concerning IT emergency management, also known as Business Continuity Management. It was released in May 2012. Its objective is to provide assistance to reduce business interruptions caused by unforeseen emergencies. Principally it is an enhancement of the standards ISO 31000 [111] and ISO 27001 [112]. It is universal in the sense that it is applicable to businesses of any size and independent of the technologies employed.

The ISO 22301 standard makes demands on organisations, requires essential analytical preparation, detailed planning, and defines responsibilities within certain designated areas.

The ISO 22301 represents a real set of rules for the build up and documentation of an emergency management with the objectives.

- To organise ways and means for adequate reactions in preparation to any emergency situation
- To quickly resume business processes
  - Firstly by providing for operations under an emergency
  - Secondly by providing a systematical approach for the resumption of all processes after the end of an emergency.

- Avoidance of emergencies by relevant preventive measures
- Minimisation of damages, again by relevant preventive measures.

Historically this means a trend change away from emergency planning to emergency management, that is, the conceptualisation of a management system in its own right.

9.2.2 Further Standards and Methodologies Concerning IT Security

There are the 2700x series, although presenting basic guidelines and recommendations, but lagging behind the details. After ISO 22301 it can be regarded as obsolete.

Basically any methodology (ITIL [113] and others) contain some references, sometimes whole sections, about the subject of emergency management. People using such methodologies, and if these have already been introduced in an organisation, should initially check there, whether an emergency management could meaningful be built upon these. The measures proposed there could possibly be supplemented by elements from ISO 22301.

9.3 Requirements for Businesses

Emergency management systems are integrated management systems and not just a collection of rules. Therefore, it is imperative that the top level management of an organisation is not only embedded in the process but has to push the project. This means to provide for personnel and material resources. Separate responsibilities have to be allocated for monitoring under regular review. They have to report to management.

The top level management should equally be involved regarding business continuity management—the continuation of the most important business processes, because in the end only they can determine priorities. On the one hand, an emergency management is of prime importance for the existence of an organisation, on the other hand it is also a demonstration to external observers to ensure that business partners and clients maintain their confidence that the company is in a position to continue its business at all. For this reason emergency management systems can be certified.

Just as with other management systems emergency management is subject to some kind of Deming Cycle, also known as PDCA Cycle (plan-do-check-act) (Fig. 9.1). Everything starts with a plan. This includes:

- Organisational structures
- Management responsibilities
- Process planning
- Resources

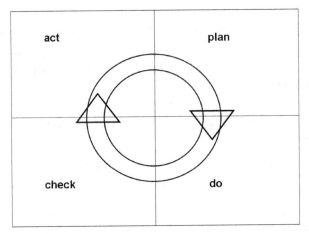

Fig. 9.1. PCDA Cycle.

Doing means the execution of the following tasks:

- Business impact analysis
- Risk analysis
- Business continuation strategy
- Tests and Training

  Checking

- Means the evaluation of the measures to be introduced
- Is the basis for internal audits
- Serves as management review

The results contain consequences for further action and are the virtual trigger for the PCDA cycle.

9.3.1 Analysis before Planning

Before detailed planning of emergency management a careful analysis about the main strategic orientation of the business has to be done, and how other basic conditions are represented in the organisation. This includes the recording of all business activities as they are for example represented in ERP processes.

Of importance are interfaces to other organisational elements, which exist already, for example to a risk management already in place, which overlap with an emergency management. Care has to be taken of the expectations of all participants of the overall process as well as attention to legal requirements.

9.3.2 Management Responsibilities

Top level management has to make sure that business continuity management (BCM) does not become a foreign matter within the organisation, but fits the overall strategy. This means integrating BCM processes into the existing process landscape. This is best achieved by formulating a BC strategy respecting the following aspects:

- Documentation of objectives and emergency plans: summarize those business activities absolutely necessary in an emergency with respect to strategic objectives to keep the business going; emergency plans have to consider this.
- Provide the necessary communication structures: communication under an emergency differs significantly from that in normal operations. The relevant channels have to be defined.
- Definition of responsibilities: for a functioning emergency organisation proper hierarchies have to be created.

Such a first formulated strategy with its associated objectives is not something static, but exists together with changes of the business operation and thus is subject to regular review and control.

9.4 BCM Overview

9.4.1 Phases and Steps of BCM Realisation

The phases of BCM realisation are displayed schematically in Fig. 9.2.

A BIA (Business Impact Analysis, s. section below) is followed by risk assessment by experts and the business leadership. On this basis the BCS (Business Continuity Strategy) is developed as well as the procedures to assure business operations under emergencies. Once these concepts have been completed training handbooks have to be written. According to these

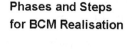

Phases and Steps for BCM Realisation

Fig. 9.2. BCM Realisation.

guidelines emergencies can be trained in full or partially and the procedures be brought to a test.

9.4.2 Business Impact Analysis (BIA)

Business Impact Analysis is an elaborate undertaking, which—as a preventive measure—may commit important company resources punctual: specialists, members of top management, executives. This analysis comprises:

- Collection and identification of processes und functions, including all procedures, not only the critical ones or those belonging to the core processes
- Underlying resources: personnel, hardware resources as IT infrastructure, buildings, stores with their technical equipment
- Dependencies on IT processes; the critical question in this case concerns those process parts, which could be carried out meaningfully without direct IT support.
- Priorities: not later than now the decision about core processes has to be taken.
- Impact and re-start-up durations; impact scenarios obviously vary with the type of assumed emergency and thus the projected re-start-up durations.

All these are preconditions to be able to execute a risk analysis and evaluation.

9.4.2.1 BIA and Risks

In sum the Business Impact Analysis

- Is a method to identify critical business processes
- Determines the impact of process breakdowns
- Shows the dependencies among processes
- And generates the required re-start-up durations.

BIA and risk analysis are for all intents and purposes the backbone of emergency management. They:

- Are the basis fort he complete emergency concept
- Define, what an emergency is
- Identify the context and threats

The awareness that processes in a company are logically interlinked and that nearly no part of the business can get along without IT processes, determines the need to document these processes under a BIA to identify the critical systems afterwards (Fig. 9.3)

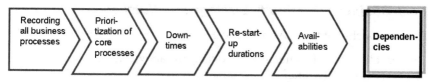

Fig. 9.3. BIA Steps.

When evaluating these processes certain aspects have to be taken into account. They cover:

- All impacts (logistical, finacial, legal, etc.)
- Obstructions or impossibilities to carry out tasks and activities
- Internal image damage and in the market, and finally
- Life and limb of the staff.

Risk Analysis

Figure 9.4 shows the procedure for risk analysis schematically. Only after BIA and the determination of critical processes the real threat can be identified and thus emergency planning can commence.

However, nothing has been said about acceptance and tolerance concerning a known risk. These are determined by other aspects:

- Endangerment of the strategy
- Costs
- Readiness by top management to assume a risk
- Practical possibilities for avoidance.

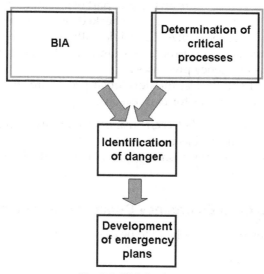

Fig. 9.4. Risk Analysis.

9.4.3 Business Continuity Strategy in a Nutshell

The relationship of the BIA and critical processes is summarised in Fig. 9.5.

- Development of strategies considering the above mentioned (readiness to assume risks, etc.)
- Identification of measures: the BIA results limit options for action to the absolute necessary
- Protection of critical activities and later re-construction in the context of intended re-start-up durations within defined objectives given by the emergency strategy
- Integral part of the business strategy: emergency strategy must not contradict the overall business strategy on the time scale.

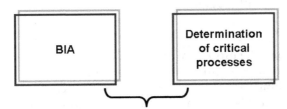

Fig. 9.5. Interplay of BIA und Process Identification.

9.5 Training and Testing

The Norm 22301 requires regular tests and training. Only tests can show how realistic theoretical assumptions for re-start-up durations are for example. In this way training concepts are an integral part of an emergency handbook. During training phases it should be checked how the performance of emergency processes looks like, if possible with quantifiable time targets. At the same time standard processes from normal operations should be compared with emergency processes.

One can and should test the developed procedures even outside an acute emergency—and this for two reasons:

- To assure consistency between business continuity management with business continuity objectives
- To make sure the chosen strategies will deliver the best answers and recovery results.

9.6 Contents of an Emergency Concept (Documentation)

A number of key documents have to be developed:

- Emergency plans

This is a detailed process documentation describing all procedures from the time of direct triggering off of the emergency alarm until recovery of normal operations—with all relevant organisations and responsibilities as well as time tables.

- Guideline
- Handbook

But before all these documents can be established the overall emergency concept has to be developed. This master document contains all the results from the analytical phase. These comprise:

- The main IT supported processes: the core processes of the business
- Results of the impact analysis
- Time table for re-start
- A rough outline for an emergency organisation (responsible emergency manager, emergency coordinators, emergency task force)
- Criteria for emergency definition (incident, emergency, crisis, catastrophe)
- Precautionary measures in addition to the emergency organisation: technical and logistical redundancies.

According to the above emergency management distinguishes between: the two aspects:

- Emergency prevention and
- Emergency mastering.

9.6.1 Guideline

The emergency management guideline should contain the following aspects:

- Definition of emergency management
  - Importance for the own organisation
  - Responsibilities
  - Interplay with other parts of the business.

    ◆ **Chosen Procedural Modell**
       – in this case *ISO Standard 22301*

    ◆ **Commitment to optimize Emergency Management**

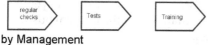

by Management

Fig. 9.6. Responsibilities of Top Management.

- Scope of the emergency management
  - Divisions
  - Objects
  - Locations
  - Duration of validity

- Compatibility with other business objectives
- Main aspects of the emergency strategy:
  - Selected strategic objectives
  - Threat scenarios
  - Readiness to assume certain risks
  - Damage scenarios
  - Priorities within business operations

- Process model (*ISO Standard 22301*)
- Secure emergency functions (Fig. 9.6)
- Legal frameworks
- Formal assumption of responsibility through signatures by the top management.

9.7 Intermediate Conclusions

In addition to strategic considerations, which are largely aimed at well defined objectives to be met by specific measures, the relevant procedures are to all intents and purposes the associated small print. First of all they contain communication guidelines within the emergency organisation, but also to external addressees, which have to function immediately in case of an emergency. Immediate actions have to be executed instantaneously to avoid acute dangers—even in the case of threats, which could not have been foreseen. But also hypothetical scenarios and their impacts on peripheral processes have to be developed to limit damage.

Emergency strategy fixes minimal objectives essential for meaningful business operations. They comprise:

- Definition of the restoration duration für critical activities deduced from the core processes
- Early availability: deduced from restoration duration; if their results prove to be inacceptable, one has to develop alternative strategies until useful targets are reached.
- Orientation along the overall business strategy and therefore integral part of the company's strategy as such although emergency strategy – obviously it has to be mapped on the original business strategy albeit in a reduced manner.

All these mentioned considerations have a superior aim as such:

> The organisation has to document procedures to ensure the continuity of activities and to come to terms with the interruption of operations!

9.7.1 Continuation of Business Processes

As already mentioned, emergency plans serve to continue business operations, even in a reduced manner. These plans must contain:

- First steps after crises and emergencies
- Instructions how to launch alternative solutions.

The following aspects have to be observed:

- Scope (which organisational units, which seasons, for which locations and perhaps which affiliated companies)
- Responsibilities at all levels and within the emergency organisation itself
- Involved persons (to be quoted by name for the members of the emergency organisation, otherwise by position in the structural organisation of the company)
- Paths of escalation (within the emergency organisation, to top management, externally to relief agencies)
- Trigger criteria for the start and finish of emergency operations (From when does one talk about "emergency", which conditions have to be fulfilled for all-clear—without taking possible repair and restoring works into consideration?)
- Repair and restoring works: one has to bear in mind that after an emergency normal business operation may have resumed, but quite often the same resources have to deal with repair and restoring work. In some cases one has to consider, if certain activities may not be executed by temporal additional capacities.
- Re-start-up plans play a particular role. They comprise:
  - Steps for re-start-up (correct sequence because of dependencies between processes among themselves)
  - Steps for restoration (re-start-up is not the same as restoration; restoration may for example include the reconstruction of destroyed buildings)
- Cycle: error correction, commencement of emergency operations, start-up of transitional solutions, return to normal operations (Fig. 9.7)

9.7.2 Step Sequence

Step 1:   Selection of standard
Step 2:   What does Business Continuity Management mean for the company in
              relation to its core processes?
Step 3:   Execution of Business Impact Analysis
Step 4:   Preparation of tests and training
Step 5:   Consolidation of emergency concept

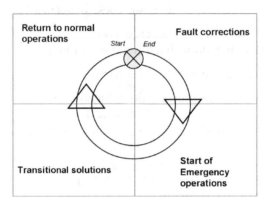

Fig. 9.7. Emergency Cycle.

9.8 The Emergency Process

The exact timing for the initiation of the emergency process, having been planned long ago, is decisive for its success once an emergency occurs. This implies the assignment of the responsibility, which person shall or must initiate the process. Normally top management is responsible for that. In case they are no longer functioning the plans should already contain relevant safety steps to allow for other authorised instances to take over the task. However, for our further considerations it is assumed that top management is still functioning. In this case its responsibilities are:

- Initiation
  To declare officially that there is an emergency—and at the same time already stating, which category is assumed (incidence, emergency, crisis, etc.). In case of "emergency" information about limitations should be given, if possible (locations, facilities, victims, etc.)

- Manage
  Although the roles of the emergency commissioner, the crisis task force, and the emergency coordinators have been fixed beforehand, the overall direction should still reside at the top level.

- Control
  By using the reporting paths previously defined top management remains up-to-date and can intervene in the emergency process if necessary.
- Overall responsibility for:
  - Resources
  - Financial means.

This means practically:

- Members of top management are the *owners of the emergency process.*
- They delegate to *the emergency commissioner and subordinated organisational units.*

9.8.1 Conception and Planning

Figure 9.8 shows a rough outline of the most important steps to create an emergency concept. The following is apparent:

- The emergency management is a process.

Emergency prophylaxis cannot be handled by a small group or commissioned staff sort of beside daily business, but has to be initiated and conducted like an elaborate project—with all organisational means and tools required by proper project management (milestone plan, project organisation, etc.).

- Objectives have to be defined.

Fig. 9.8. Development of an Emergency Concept.

These include:

- o Planning of time scales and resources (Who can be recruited from which divisions in the company, how much expenditure of time has to be budgeted, what special competences are required?)
- o Definition of Scope (locations, organisational units, affiliated companies)
- o Establish a general framework (premises, exemptions from daily business, IT infrastructure, communication equipment)
- o Strategy definition.

Strategy definition is deduced from core processes, risk analysis, and occurrence probability.

9.8.1.1 Scope

The question about scope for an emergency concept has to be viewed from different angles:

- Is the whole institution concerned or only individual locations or departments (being critical for the continued functioning of the institution as such)?
- Are all processes concerned or only certain ones (critical ones, core processes)?
- Do limitations or boundaries exist; do they have to be documented and justified?
- Even, if all processes have to be included the most important ones have to be highlighted with a particular weight.

All these considerations have to respect legal requirements. Table 9.1 cites some important laws and is far from complete. These laws may differ from country to country.

Table 9.1. Legal Requirements.

Sarbanes Oxley Act [114]
Basel Capital Accord [115]
Stock Corporation Acts, Security Investor Protection Act
Laws Regulating Postal and Telecommunication Services
Stock Exchange Acts
Laws Regulating Occupational Safety
Hazardous Incident Ordinances
Ordinances on Hazardous Substances
Laws on Operational Safety

9.8.1.2 Other Requirements

Basically one has to decide upon those business objectives, which are indispensable and have to be reached in spite of an emergency. With this in mind, possible damage scenarios have to be imagined, which could impede these objectives. These considerations lead to simulations concerning the interruption of critical processes, which may impede the further functioning of an organisation (Fig. 9.9).

Furthermore, these considerations form the basis for assessing which risk the company is going to accept and which measures are necessary to avoid the passing of a risk borderline. This will lead to the conclusion about which practical objectives the emergency process wants to head for. From this the particular interest of stakeholders are affected. Potential stakeholders are:

- Shareholders
- Staff
- Relatives
- Investors
- Clients
- Suppliers
- Insurers
- Regulating authority
- Trade organisations
- Legislators.

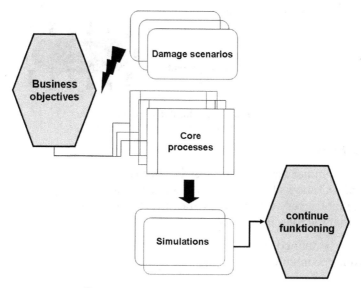

Fig. 9.9. Retention of Business Objectives.

The organisational requirements to overcome an emergency fall into the following categories:

- Emergency prevention and
- Emergency mastering.

These two aspects themselves operate on three levels:

- Strategic
- Tactical
- Operational.

Figure 9.10 shows the schematic distribution of roles and responsibilities as listed in Table 9.2.

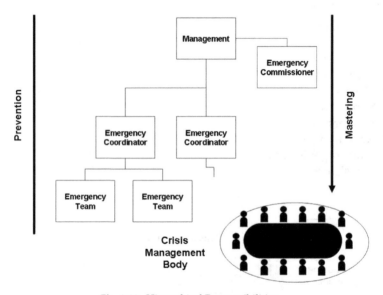

Fig. 9.10. Hierarchical Responsibilities.

9.8.1.3 Roles

Table 9.2 summarizes roles and responsibilities.

Table 9.3 shows the tasks of the emergency teams.

The tables very clearly show the enormous coordination effort. This means that the roles have to be:

- Clearly defined and
- Well documented.

Table 9.2. Roles and Responsibilities.

Management	Ensure emergency management all over the institution
Emergency Commissioner	Controls all activities in connection with emergency prevention
	Coordinates the allocation of resources
	Analyses the overall process of emergency mastering after a damaging incident
	Appoints and manages those with responsibility for the implementation of necessary measures
	Reports to management
	Needs to have the qualifications essential for his job
Emergency Coordinators	Support the Emergency Commissioner for every major logical unit of an institution
	Interface between Emergency Commissioner and organisational unit
	Participation in tests and training
	Analysis of results
	Report to the Emergency Commissioner
Emergency Prevention Team	Selected experts
	Consult the Emergency Commissioner and the Emergency Coordinators
Crisis Management Body	Strategic responsibility and decisions
	Representing members of management
	Liaison with stakeholders
Emergency Task Force	Plans, coordinates, consults, supports
	Determination of the current situation and evaluation
	Issuing of tasks to responsible instances
	Coordination of necessary activities
	Crisis communication
	Clearing of individual measures
Head and core team of the Emergency Task Force	One head and up to five function owners
	With deputy
	Residing locally
Task Force Members	Representative for public relations
	Security manager
	Member of IT operations
	Necessary competence and experience
Extended Emergency Task Force	IT administration/IT manager
	Local safety (fire protection, environmental protection, plant security, rescue service)
	Legal adviser
	Employee representation

Table 9.2 contd....

...Table 9.2 contd.

	Contact person for departments
	Data privacy protection representative
Supporting additional resources (for example Company Medical Officer)	
External specialists	
Emergency Teams	Operational mastering
	Emergency team leader reports to the Emergency Task Force

Table 9.3. Tasks of the Emergency Teams.

Infrastructure Team	Restoration of buildings and workplaces
IT Team	Recovery of data
	Allocation of redundant systems
	Repair disruptions of communications equipment
Departments Teams	Re-start business processes

Each qualified participant not only has to understand how he has to act in praxis of an emergency, but he also has to know, with whom he has to communicate, what his lines of communications look like, and what escalation paths are available.

9.8.2 Step Sequence

Step 1: Determine emergency notification process
Step 2: Initiation of emergency prevention project
Step 3: Determine scope
Step 4: Determine objectives and requirements for emergency operations
Step 5: Define roles and responsibilities

10

References

[1] http://searchmobilecomputing.techtarget.com/definition/wireless-LAN.

[2] IEEE Std 802.16-2009 (Revision of IEEE Std 802.16-2004), pp. C1–2004 (2009a), 29 May 2009.

[3] Code of Federal Regulations (Revised as of 1 October 2001) Title 47, vol. 1: 733–735: Telecommunication, Chapter I: Federal Communications Commission, Part 15–Radio Frequency Devices.

[4] IEEE 802.11-1997: Wireless LAN medium access control (MAC) and physical layer (PHY) specifications.

[5] IEEE 802.11: Wireless LAN medium access control (MAC) and physical layer (PHY) specifications. (2007 revision). IEEE-SA. 12 June 2007.

[6] IEEE 802.11k-2008—Amendment 1: Radio resource measurement of wireless LANs. IEEE-SA. 12 June 2008.

[7] IEEE 802.11r-2008—Amendment 2: Fast basic service set (BSS) transition. IEEE-SA. 15 July 2008.

[8] IEEE 802.11y-2008—Amendment 3: 3650–3700 MHz operation in USA. IEEE-SA. 6 November 2008.

[9] Zimmermann, H. OSI reference model—the ISO model of architecture for open systems interconnection. IEEE Transactions on Communications, COM-28(4): 425–432, April 1980.

[10] Abouzeid, A. Wireless ad hoc and Sensor Networks MAC Layer Introduction & the IEEE802.11 standard, ECSE, RPI, 15 September 2005.

[11] IEEE 802.2-1998 (ISO/IEC 8802-2: 1998), IEEE standard for Information technology —Telecommunications and information exchange between systems—Local and metropolitan area networks—Specific requirements—Part 2: Logical Link Control.

[12] Lee, B.G. and S. Choi. 2008. Broadband wireless access and local networks: mobile WiMax and Wi-Fi. Artech House, Boston, MA.

[13] Federal Standard 1037C.

[14] Popovski, P., H. Yomo and R. Prasad. 2006. Strategies for adaptive frequency hopping in the unlicensed bands. Aalborg University, IEEE Wireless Communications, December 2006.

[15] Boer, J. 1996. Direct Sequence spread spectrum physical layer specification IEEE 802.11. Lucent Technologies WCND Utrecht.

[16] Coleri, S. Ergen, M. Puri and A. Bahai. 2002. Channel estimation techniques based on pilot arrangement in OFDM systems. IEEE Transactions on Broadcasting, September 2002.

[17] http://www.intel.com/products/centrino/

[18] 802.11i-2004. IEEE standard for information technology—telecommunications and information exchange between systems—Local and metropolitan area networks—Specific requirements, Part 11: Wireless LAN Medium Access Control (MAC) and Physical Layer (PHY) specifications, Amendment 6: Medium Access Control (MAC), Security Enhancements, 2004.

[19] http://www.wi-fi.org/

[20] Kozierok, C.M. 2005. The TCP/IP guide: A comprehensive, illustrated internet protocols reference. No Starch Press, San Francisco.

[21] National Institute of Standards and Technology (NIST), Advanced Encryption Standard (AES). FIPS 197, 26 November 2001.

[22] The Internet Society. 2004. Network Working Group, RFC 3748, Extensible Authentication Protocol (EAP).

[23] The Internet Society. 2000. Network Working Group, RFC 2865, Remote Authentication Dial In User Service (RADIUS).

[24] Studium, P. and A.S. Tanenbaum. Computernetzwerke, p. 268ff.

[25] Specification of WAP conformance requirements. WAP Forum, WAP-221-CREQ-20010425-a. Available at http//www.wapforum.org/.

[26] IEEE-Report: Status of Project IEEE 802.11n

[27] IEEE P802.11. Task Group p - Meeting Update, Status of project IEEE 802.11 Task Group p, 2010.

[28] Tse, D. and P. Viswanath. 2005. Fundamentals of wireless communication. Cambridge University Press.

[29] Johnsson, M. 1999. HiperLAN/Z—The broadband radio transmission technology operating in the 5 GHZ frequency band. HiperLAN/2 Global Forum.

[30] Dean, T. 2004. Network + Guide to Networks 3rd Edition. Course Technology, Boston, MA.

[31] Fazel, K. and S. Kaiser. 2008. Multi-carrier and spread spectrum systems: from OFDM and MC-CDMA to LTE and WiMAX, 2nd Edition. John Wiley & Sons.

[32] Egevang, K. and P. Francis. 1994. The IP network address translator (NAT), RFC 1631, May 1994.

[33] Authorization of spread spectrum systems under parts 15 and 90 of the FCC rules and regulation (TXT), Federal Communications Commission. 18 June 1985. Retrieved from http://www.marcus-spectrum.com/documents/81413RO.txt

[34] IEEE Std 802.11-2007. Section 3.16, p. 6. 12 June 2007.

[35] Microsoft. 2003. Microsoft TechNet. Retrieved from How 802.11 wireless works: http://technet.microsoft.com/enus/library/cc757419%28WS.10%29.aspx.

[36] Ohrtman, F. and K. Roeder. 2003. Wi-Fi handbook: building 802.11b wireless networks. McGraw Hill, New York.

[37] http://www.tech-faq.com/ssid.html

[38] https://support.microsoft.com/en-us/help/310563/description-of-internet-connection-sharing-in-windows-xp

[39] Roy, R.S. and B. Ottersten. 1996. Spatial division multiple access (SDMA), US Patent No. 5515378.

[40] Samartini, P. and S. Capitani di Vimercati. 2001. Access control: policies, models, and mechanisms. *In*: Foundations of Security Analysis and Design: Tutorial Lectures, Lecture Notes in Computer Science 2171: 137–193.

[41] Vladimirov, A., K.V. Gavrilenko and A.A. Mikhailovsky. 2004. Wi-Foo: The Secrets of Wireless Hacking. Pearson/Addison Wesley, US.

[42] Aycock, J. 2010. Spyware and adware (Advances in Information Security). Springer, New York.

[43] What is the difference: viruses, worms, trojans, and bots? (n.d.). Cisco. Retrieved 22 September 2014, from http://www.cisco.com/web/about/security/intelligence/virus-worm-diffs.html.

[44] Connolly, K.J. 2003. Law of internet security and privacy. pp. 131, Aspen Publishers, US.

[45] Spitzner, L. 2002. Honeypots tracking hackers. Addison Wesley, US.

[46] Mäusl, R. and J. Göbel. 2002. Analoge und digitale Modulationsverfahren. 1. Auflage Hüthig, Germany.

[47] IEEE Std 802.11b-1999, §18.4.6.5.

[48] Lathi, B.P. 1983. Modern digital and analog communication systems. Hault-Saunders, New York.

[49] Lee, B.G. and S. Choi. 2008. Broadband wireless access and local networks: mobile WiMax and Wi-Fi. Artech House, Boston, MA.

[50] Peterson, W.W. and D.T. Brown. 1961. Cyclic Codes for Error Detection. *In:* Proceedings of the IRE 49, p. 228, January 1961.

[51] Federal Standard 1037C.

[52] 802.11i-2003. IEEE Standard for Information technology—Telecommunications and information exchange between systems—Local and metropolitan area networks—Specific requirements, Part 11: Wireless LAN Medium Access Control (MAC) and Physical Layer (PHY) specifications, Amendment 6: Medium Access Control (MAC), Security Enhancements, 2003.

[53] Tanenbaum, A.S. 2003. Computernetwerken (Computer Networks), fourth edition Ed. Pearson Prentice Hill, US.

[54] Plummer, D.C. 1982. RFC 826, An Ethernet address resolution protocol—or—converting network protocol addresses to 48.bit Ethernet address for transmission on Ethernet hardware. Internet Engineering Task Force, Network Working Group, 1982.

[55] Gantz, J. and J.B. Rochester. 2005. Pirates of the digital millennium. FT Prentice Hall, Upper Saddle River, NJ.

[56] Wireless Networking Basics, NETGEAR, Inc. 4500 Great America Parkway Santa Clara, CA 95054 USA, 2005.

[57] airsnort.shmoo.com

[58] Robshaw, M. 2006. Fast software encryption: 13th International Workshop. FSE 2006, Graz, Austria, 15–17 March 2006, Revised Selected Papers (Lecture Notes in Computer Science), Springer, 23 August 2006.

[59] Sobh, T. et al. 2008. Novel algorithms and techniques in telecommunications. *In:* Automation and Industrial Electronics. Springer.

[60] Wi-Fi alliance announces standards-based security solution to replace WEP. Wi-Fi Alliance. 31 October 2002.

[61] IEEE Standard 802.1X-2004—Port-based network access control.

[62] The Internet Society. 2004. Network Working Group, RFC 2865, Extensible Authentication Protocol (EAP).

[63] www.openseaalliance.org

[64] US allows spectrum use to speed up wireless communications. Computer, IEEE Computer Society, November 2010.

[65] Paulson, L.D. 2011. Wireless devices provide users with mobile Wi-Fi hotspots. Computer, IEEE Computer Society, January 2011.

[66] AVM Computersysteme Vertriebs GmbH, Berlin. Retrieved March 2011.

[67] 3GPP TS 11.11. Specification of the subscriber identity module - mobile equipment (SIM-ME) interface. Retrieved from https://portal.3gpp.org/desktopmodules/Specifications/SpecificationDetails.aspx?specificationId=419.

[68] Becker, A. et al. 2010. Android 2. Grundlagen und programmierung, dpunkt. Verlag, Heidelberg.

[69] Williams, R. 2015. Apple iOS: a brief history. The Telegraph, 17 September 2015.

[70]   Alastair, S. 2009. BlackBerry planet: the story of research in motion and the little device that took the world by storm. John Wiley & Sons, Canada.

[71]   BlackBerry mobile data system. Retrieved from http://us.blackberry.com/apps-software/mobile.jsp.

[72]   Machat, P. 2008. Die BlackBerry-Sicherheitsarchitektur aus der Nähe betrachtet. RIM.

[73]   http://www.gsmworld.com/about-us/history.htm

[74]   ETSI EN 301 349 V8.4.1 (2000–10).

[75]   Collins, D. and C. Smith. 2001. 3G wireless networks. McGraw-Hill.

[76]   Haid, M. 2001. HSCSD als Leistungsmerkmal im GSM-Mobilfunk der Generation 2+. Hagen.

[77]   Sauter, M. 2006. Communication systems for the mobile information society. John Wiley, Chichester.

[78]   GSM Doc 28/85. 1985. Services and facilities to be provided in the GSM system, rev2, June 1985.

[79]   WAP Architecture. 2001. Wireless application protocol architecture specification WAP-210-WAPArch-20010712. Wireless Application Protocol Forum. Available at http://www1.wapforum.org/tech7docurnentsAVAP-210-WAPArch-20010712-a.pdf, last accessed in July 2001.

[80]   Vacca, J.R. 2001. I-mode crash course. McGraw-Hill.

[81]   Roth, J. 2005. Mobile computing. Grundlagen, Technik, Konzepte, Dpunkt, Heidelberg.

[82]   http://www.umatechnology.org/overview/index.htm

[83]   RFC 3851. Secure/multipurpose Internet mail extensions (S/MIME) version 3.1 message specification.

[84]   Hendry, M. 2007. Multi-application smart cards. Cambridge University Press, UK.

[85]   Whitfield, D. and M. Hellman. 1976. New directions in cryptography. IEEE Transactions on Information Theory, IT-22z: 644–654, November 1976.

[86]   International Engineering Consortium. 2007. Voice over Internet protocol. Definition and Overview. Retrieved from http://www.iec.org/online/tutorials/ int_tele/index.asp.

[87]   Rohwer, T. et al. 2006. Abwehr von spam over internet telephony (SPIT-AL), TNG, Kiel.

[88]   http://www.bluetooth.com/Pages/Bluetooth-Home.aspx

[89]   Sven-Olaf Suhl, Bluetooth 2.1+DER verspricht simple Geräte-Kopplung, heise.de, 3 August 2010.

[90]   http://bluetooth.com/SiteCollectionDocuments/Core_V30HS.zip

[91]   https://bluetooth.com/what-is-bluetooth-technology/how-it-works/low-energy

[92]   S. Rathi. 2000. Bluetooth protocol architecture. Dedicated Systems Magazine.

[93]   Bluetooth tutorial profiles. Retrieved 5 January 2007 from http://www.palowireless.com/infotooth/tutorial/profiles.asp.

[94]   F. Bennett et al. 2011. Piconet embedded mobile networking. The Olivetti and Oracle Research Laboratory, Cambridge, UK.

[95]   https://www.bluetooth.com/specifications/bluetooth-core-specification/bluetooth5

[96]   Scarfone, K. and J. Padgette. 2008. Guide to bluetooth security: Recommendations of the National Institute of Standards and Technology. Special Publication 800-121, National Institute of Standards and Technology (NIST), US Department of Commerce.

[97]   Uzun, E. et al. 2011. Usability analysis of secure pairing methods. University of California, Irvine.

[98]   Knutson, C.D. and J.M. Brown. 2004. IrDA principles and protocols. MCL Press, Salem UT.

[99]   Deicke, F. 2007. Optische drahtlose Datenübertragung. Fraunhofer IPMS.

[100]   Palmer, M. 2004. Wireless communication using the IrDA® standard protocol. Microchip Technology Incorporated.

[101]   Yeh, K.W. and L. Wang. 2011. An introduction to the IrDA standard and system implementation. Hewlett & Packard.

[102] Faulkner, C. 2017. What is NFC? Everything you need to know. Retrieved from www. techradar.com.

[103] https://nfc-forum.org

[104] Ben Morris, B. 2007. The Symbian OS Architecture Sourcebook: Design and Evolution of A Mobile Phone OS. John Wiley & Sons, 2007.

[105] https://sony.net/Products/felica

[106] ISO/IEC 13335-1:2004

[107] ISO/IEC 17799:2005

[108] ISO 17799 Newsletter. 2005. News & Updates for ISO 27001 and ISO17799.

[109] Aguyao, R. 1991. Dr. Deming: The American who taught the Japanese about quality. Fireside.

[110] https://www.iso.org/standard/50038/html

[111] https://www.iso.org/iso-31000-risk-management.html

[112] https://www.iso.org/standard/54534.html

[113] https://www.axelos.com/best-practice-solutions/itil

[114] Kimmel et al. 2011. Financial Accounting. Wiley.

[115] Marrison, C. 2002. The fundamentals of risk management. McGraw Hill, New York.

Index